CATALYST

A framework for success

Carol Chapman
Moira Sheehan

Heinemann
Inspiring generations

Contents

T indicates Think about spread

Introduction

Welcome to Catalyst

This is the first of three books designed to help you learn all the science ideas you need during Key Stage 3. We hope you'll enjoy the books as well as learning a lot from them.

This book has twelve units which each cover a different topic.
The units have two types of pages:

Learn about:

Most of the double-page spreads in a unit introduce and explain new ideas about the topic. They start with a list of these so that you can see what you are going to learn about.

Think about:

Each unit has a double-page spread called Think about. You will work in pairs or small groups and discuss your answers to the questions. These pages will help you understand how scientists work and how ideas about science develop.

On the pages there are these symbols:

ⓐ   Quick questions scattered through the pages help you check your knowledge and understanding of the ideas as you go along, for example,

ⓐ   **Use the particle model to explain why the liquid will not squash .**

Questions

The questions at the end of the spread help you check you understand all the important ideas.

For your notes:

These list the important ideas from the spread to help you learn, write notes and revise.

Do you remember?

These remind you of what you already know about the topic.

Did you know?

These tell you interesting or unusual things, such as the history of some science inventions and ideas.

At the back of the book:

Glossary

All the important scientific words in the text appear in bold type. They are listed with their meanings in the Glossary at the back of the book. Look there to remind yourself what they mean.

Index

There is an Index at the very back of the book, where you can find out which pages cover a particular topic.

Activities to help or check your learning:

Your teacher may give you these activities from the teacher's materials which go with the course:

Transition quiz and worksheets

Before you start a new unit your teacher may give you a quiz or a worksheet. This short exercise will help you remember what you already know about a topic.

Unit map

You can use this to think about what you already know about a topic. You can also use it to revise a topic before a test or exam.

Starters

When you start a lesson this is a short activity to introduce what you are going to learn about.

Activity

There are different types of activities, including investigations, that your teacher can give you to help with the topics in each spread in the pupil book.

Plenaries

At the end of a lesson your teacher may give you a short activity to summarise what you have learnt.

Homework

At the end of a lesson the teacher may give you one of the homework sheets that go with the lesson. This will help you to review and revise what you learnt in the lesson.

Pupil check list

This is a check list of what you should have learnt to help you with your revision.

Test yourself

You can use this quiz at the end of each unit to see what you are good at and what you might need to revise.

End of unit test Tier 3-6

This helps you and your teacher check what you learnt during the unit, and measures your progress and success.

Heinemann Educational Publishers
Halley Court, Jordan Hill, Oxford OX2 8EJ
Part of Harcourt Education

Heinemann is the registered trademark of
Harcourt Education Limited

© Carol Chapman, Moira Sheehan 2003

First published 2003

07 06 05 04 03
10 9 8 7 6 5 4 3 2 1

British Library Cataloguing in Publication Data is available
from the British Library on request.

ISBN 0 435 76010 6

Edited by Ruth Holmes and Sarah Ware
Designed by Ken Vail Graphic Design
Typeset by Ken Vail Graphic Design

Original illustrations © Harcourt Education Limited 2003

Illustrated by Graham-Cameron Illustration (Tim Archbold, Darin Mount and Sarah
Wimperis), Nick Hawken, Illustration Ltd (David Ashby), B. L. Kearley (Sheila Galbreath,
Jeremy Gower and Pat Tourett), Linda Rogers Associates (Lorna Barnard, Dave
Burroughs, Keith Howard, Gary Rees and Branwen Thomas), David Lock, Joseph McEwan,
John Plumb, Simon Girling & Associates (Mike Lacey), Sylvie Poggio Artists Agency (Nigel
Kitching, Rhiannon Powell, Lisa Smith and Sean Victory)

Printed in the UK by Bath Press Ltd

Picture research by Jennifer Johnson

Acknowledgements
The authors and publishers would like to thank the following for permission to use
copyright material: **bar chart p92**, Addison-Wesley, *Body Maintenance*; **graph p47**,
The Canadian Journal of Rural Medicine Vol. 3, p12–19, with permission from the Society
of Rural Physicians of Canada; **map p8**. This product includes mapping data licensed from
Ordnance Survey® with permission of the Controller of her Majesty's Stationary Office, ©
Crown copyright. All rights reserved. License no. 100000230.

The publishers have made every effort to trace the copyright holders, but if they have
inadvertently overlooked any, they will be pleased to make the necessary
arrangements at the first opportunity.

For photograph acknowledgements, please see page vii.

Tel: 01865 888058 www.heinemann.co.uk

The author and publishers would like to thank the following for permission to use photographs:

T = top   **B** = bottom   **L** = left   **R** = right   **M** = middle

SPL = Science Photo Library

Cover: Getty Images.

Page 2, **L**: Biophoto Associates; 2, **R**: SPL/M.I. Walker; 3, **T**: SPL/Eye of Science; 3, **B**: Corbis; 5: SPL/A.B. Dowsett; 6, **T**: SPL/M.I. Walker; 6, **B** x3: Biophoto Associates; 7, x3: Biophoto Associates; 10: SPL/Andrew Syred; 12: Bruce Coleman; 14, **T**: SPL/Science Pictures; 14, **B**: SPL/Andy Walker; 15, **L**: SPL/D. Phillips; 15, **R**: SPL/Don Fawcett; 16, **TL**: SPL/Dr. Yorgos Nikas; 16, **TR**: SPL/Dr. Yorgos Nikas; 16, **B1**: SPL/Profs PM Motta & S Makeabe; 16, **B2**: SPL/Petit Format, Nestle; 16, **B3**: SPL/Alex Bartel; 16, **B4**: S&R Greenhill; 19, **T**: S&R Greenhill; 19, **B**: S&R Greenhill; 20: S&R Greenhill; 22: Bruce Coleman; 22, inset: NHPA/A.N.T.; 23, **T**: NHPA/A Warburton & S. Toon; 23, **B**: Bruce Coleman; 24, **T**: SPL/Tony Craddock; 24, **M**: SPL/Tom McHugh; 24, **B**: Hilary Fletcher; 25, **TL**: FLPA/A Christiansen; 25, **TR**: FLPA/David Hosking; 25, **M**: FLPA/Philip Perry; 25, **B**: FLPA/Jurgen & Christine Sohns; 26, **T**: Heather Angel; 26, **B**: Heather Angel; 27: FLPA/D P Wilson; 28, **TL**: FLPA/David Hosking; 28, **TR**: FLPA/M J Thomas; 28, **ML**: Holt Studios/Nigel Cattlin; 28, **MR**: Holt Studios/Nigel Cattlin; 28, **B**: OSF/Tim Shepherd; 29, **T**: FLPA/Silvestris; 29, **BL**: Corbis/Steve Kaufman; 29, **BR**: SPL/Brock May; 30, **TL**: Bruce Coleman/ Paul van Gaalen; 30, **TR**: NHPA/Martin Harvey; 30, **BL**: SPL/Dr Jeremy Burgess; 30, **BR**: Bruce Coleman/Kim Taylor; 31, **TL**: Bruce Coleman/Joe McDonald; 31, **TR**: Bruce Coleman/ Sarah Cook; 31, **M**: Bruce Coleman/Joe McDonald; 31, **B**: NHPA/Dave Watts; 33, **T**: FPLA/Images of Nature/John Hawkins; 33, **B**: FLPA/David Hosking; 35, **TL**: NHPA/Roy Waller; 35, **TM**: Heather Angel; 35, **TR**: Heather Angel; 35, **ML**: Heather Angel; 35, **MM**: Heather Angel; 35, **MR**: FLPA/Mike J Thomas; 35, **BL**: NHPA/Stephen Dalton; 35, **BM**: NHPA/Martin Harvey; 35, **BR**: Heather Angel; 36, **L**: NHPA/Darek Karp; 36, **R**: OSF/Johnny Johnson; 37: Tony Stone; 40, **LT**: Still Pictures/Ron Gilling; 40, **LM**: Still Pictures/Francois Gilson; 40, **LB**: OSF/Harry Taylor; 40, **M1**: Hans Reinhard; 40, **M2**: OSF/John Downer; 40, **M3**: OSF/Scott Camazine/CDC; 40, **M4**: SPL/Dr Linda Stannard; 40, **RT**: OSF/David M Dennis; 40, **RB**: Bruce Coleman; 41: MEPL; 42, **T1**: Photodisc; 42, **T2**: Photodisc; 42, **T3**: OSF/Maurice Tibbles; 42, **T4**: Photodisc; 42, **T5**: Bruce Coleman; 42, **B**: FLPA/Images of Nature/ John Hawkins; 43, **T**: OSF/Max Gibbs; 43, **M**: OSF/Souricat; 43, **B**: Bruce Coleman; 44, **TL**: OSF/Fredrik Ehren Strom; 44, **TM**: OSF/Rudie Kuiter; 44, **TR**: OSF/David Fox; 44, **ML**: Bruce Coleman; 44, MM: OSF/London Scientific Film; 44, MR: FLPA/Images of Nature/Gerard Laci; 44, **B1**: OSF/Frank Schneidermever; 44, **B2**: Bruce Coleman; 44, **B3**: Bruce Coleman/Animal Ark; 44, **B4**: OSF/Mills Tandy; 45: SPL/Sinclair Stammers; 48, **T**: Andrew Lambert; 48, **B**: Andrew Lambert; 49: Andrew Lambert; 50, x7: Andrew Lambert; 52, x2: Andrew Lambert; 53: Garden and Wildlife Matters; 56, **T**: Robert Harding; 56, **ML**: Andrew Lambert; 56, **MR**: Pete Morris; 56, **B**: Robert Harding; ; 57, **T**: Gareth Boden; 57, **M**: Peter Gould; 57, **B**: Peter Gould; 58, **T**: Andrew Lambert; 58, **M**: Andrew Lambert; 58, **B**: Peter Gould; 59: Roger Scruton; 60, **T**: Andrew Lambert; 60, **M**: Peter Gould; 60, **B**: Anthony Blake Picture Library/Tim Hill; 61, **L**: The Art Archive; 61, **R**: Source Unknown; 62: Robert Harding/Shout P. Allen; 63, **TL**: Shout; 63, **TM**: Shout; 63, **TR**: Shout; 63, **B**: Bruce Coleman/Ian Sargeant; 64, **T**: Tony Stone; 64, **B**: SPL/Jerry Mason; 65, **T**: Gareth Boden; 64, **B**: Andrew Lambert; 66 Robert Harding; 68, **T**: Ancient Art & Architecture; 68, **B**: Ancient Art & Architecture; 69, **TL**: SPL; 69, **TR**: SPL/Sheila Terry; 69, **B**: IBM Corporation, Almaden Research Centre; 71: SPL/Richard Folwell; 73, **T**: Corbis/Phil Schermeister; 73, **B**: TRIP/H Rogers; 74, **T**: Andrew Lambert; 74, **B** x4: Andrew Lambert; 75, **T** x2: Pete Morris; 75, **B**: Source Unknown; 76, x2: Peter Gould; 77, **T**: Barnaby's Picture Library; 77, **M**: Andrew Lambert; 77, **B** x3: Peter Gould; 78, **T**: Roger Scruton; 78, **M** x2: Andrew Lambert; 78, **B**: Roger Scruton; 80: Peter Gould; 81: Source Unknown; 82, x2: Peter Gould; 86, x4: Peter Gould; 87: Gareth Boden; 88, **TL**: Empics; 88, **TR**: Tony Stone; 88, **M**: J Allan Cash Ltd; 88, **B**: Robert Harding; 89: Trevor Hill; 91, x3: Alan Edwards; 98: SPL/Martin Bond; 99, **TL**: Tony Stone/Yvette Cardozo; 99, **TR**: Tony Stone/Lorne Resnick; 99, **B**: Holt Studios/Nigel Cattlin; 100, **T**: Environmental Images/David Hoffman; 100, **M**: Environmental Images/David Hoffman; 100, **B**: SPL/Martin Bond; 101, **T**: Tony Stone; 101, B: AP/Steve Holland; 106, **L**: MEPL; 106, **R**: SPL/Tek Image; 106, **B**: Image Select/ Ann Ronan; 111: SPL; 114, **T**: Photodisc; 114, **M**: Action Plus; 114, **B**: Peter Morris; 115, **L**: Fiat; 115, **L** insert: Ford; 115, **R**: Empics; 115, **R** insert: Empics; 124, **TL**: SPL/Jerry Mason; 124, **TM**: SPL/Andrew McClenaghan; 124, **TR**: SPL/Amy Trustram Eve; 124, **B**: SPL/Pekka Parviainen; 125: SPL/Space Telescope Science Institute/NASA; 126: SPL/Dr. Fred Espenak; 130, Mercury: SPL/NASA, Mehau Kulyk; 130, Venus: SPL/NASA; 130, Earth: SPL/NASA; 130, Mars: SPL/US Geological Society; 130, Jupiter: SPL/NASA; 130, Saturn: SPL/Space telescope Institute/NASA; 131, Uranus: SPL/NASA; 131, Neptune: SPL/NASA; 131, Pluto: SPL/Space Telescope Institute/NASA.

A1 Organs, cells, tissues

Organs do their job

Animals and plants are made up of **organs**. An organ carries out a **function** (a job) that is necessary to keep the animal or plant alive.

teeth – break up our food

heart – pumps the blood around the body

muscles – move the body

male and female organs – in flowers

leaves – make the plant's food

roots – anchor the plant and take in water

(a) Name one organ that is not shown in the picture, and give its function, for: (i) an animal (ii) a plant.

Looking closer

Organs in your body are made of smaller parts. You can see these using a **microscope**. A microscope **magnifies** small things, making them look much bigger.

Today we know from using microscopes that all living things, or **organisms**, are made up of units called **cells**. Look at the photos below. They show cells from an animal and from a plant. Animals and plants are **multicellular** organisms. That means they contain more than one cell. Many organisms are **unicellular**, containing only one cell.

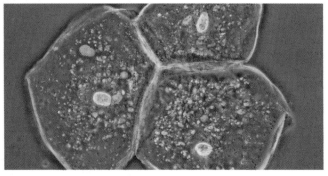

Cells from the skin inside a person's cheek.

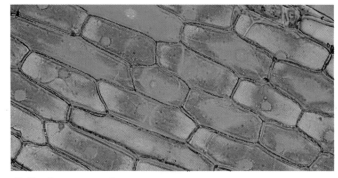

Cells from an onion.

Building an organ from cells

Think about a house. It has a roof, walls and windows. But piles of bricks, panes of glass, and roof tiles do not make a house. They have to be arranged so that the bricks make walls, the roof tiles make a roof and the panes of glass make windows.

Inside an organ, each cell has its place and similar cells work together. Look at the photo on the right. It shows a leaf cut through and then looked at using a microscope. The cells are arranged into layers. These layers are called **tissues**. The cells in one tissue are all alike. This is because they have the same function.

A group of several tissues together makes an organ. A house is rather like an organ. Its job is to give a warm, dry place in which to live. The walls and roof are like tissues, they keep out the cold and the rain. The bricks and the roof tiles are like cells, they make the walls and the roof.

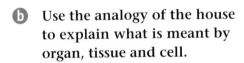

1 upper epidermis
2 palisade mesophyll layer
3 spongy mesophyll layer
4 lower epidermis

b Use the analogy of the house to explain what is meant by **organ**, **tissue** and **cell**.

In the leaf, all the **palisade cells** in the **palisade mesophyll** trap sunlight to make food.

Most animals and plants are made up of organs, which are made up of tissues, which are made up of many cells. You can see this in the human hand.

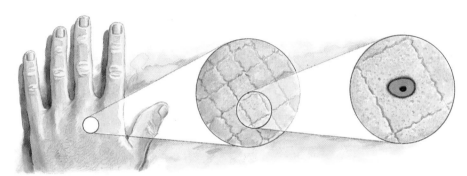

The skin is an **organ**.

The outer layer of skin is a **tissue**. It is called the **epidermis**.

This is a skin **cell**.

Cells working alone

In unicellular organisms all the life processes needed to stay alive happen inside one cell. Many algae are unicellular organisms a bit like plants. You may have seen algae growing on tree trunks. They look like green dust. The amoeba in this photo is a unicellular organism a bit like an animal. It lives in ponds and streams and feeds on tiny plants.

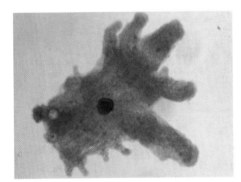

Questions

1 Write a sentence using each starting phrase.

   a An organ …   **b** A tissue …   **c** A cell …

2 Write each list below in order of size, starting with the largest each time.

   a organ, organism, cell, tissue

   b palisade cell, plant, leaf, palisade mesophyll

   c human, skin cell, epidermis, skin

3 Name: **a** seven organs and **b** five tissues mentioned on these two pages.

For your notes:

- Plants and animals are made up of **organs**. Each organ has a **function**.

- Organs are made up of **tissues**.

- Tissues are made up of **cells**. The cells in a tissue are alike.

More about cells

A cell is so small that we can see it only when we use a microscope. We say that cells are **microscopic**.

In 1665 a scientist called Robert Hooke used a microscope to look at many living things. One of the things he looked at was cork, a type of tree bark. He saw many tiny boxes. He was the first to call these boxes 'cells'.

Scientists have discovered much more about what cells look like inside, and how they work, using the **electron microscope**. Electron microscopes magnify things more than normal microscopes do.

There are two main types of cell: **animal cells** and **plant cells**. They have a lot in common, but they also have some differences.

Robert Hooke.

Animal cells

An animal cell is shown on the right. Surrounding the cell is a thin **cell membrane**. The membrane lets substances such as water and dissolved gases in and out of the cell. Inside the cell is a jelly-like substance called the **cytoplasm**. This is where **chemical changes** happen to keep the cell alive. Every cell has a **nucleus**, which controls everything that happens inside the cell.

a Why do you think the cell membrane is important?

Plant cells

A plant cell is shown on the right. It has a cell membrane, a nucleus and cytoplasm, just as animal cells do.

Plant cells also have some parts that animal cells don't have. Many plant cells have lots of small structures called **chloroplasts**. These contain a green substance called **chlorophyll**. This is why plants look green. Plants make their food in these chloroplasts

Every plant cell has a **cell wall** made of a tough, stringy substance called **cellulose**. This supports the cell, making it strong and giving it its shape.

Inside the cell is a large **vacuole**. This contains a liquid called sap that keeps the cell firm.

b Which two parts of a plant cell make it strong and firm like a tiny brick?

cell membrane

nucleus

cytoplasm

An animal cell.

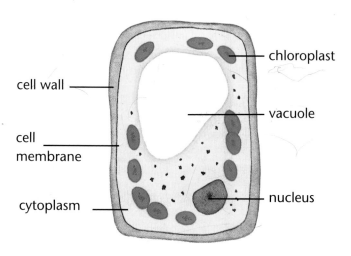

chloroplast

cell wall

vacuole

cell membrane

cytoplasm

nucleus

A plant cell.

Different cells

The life processes that multicellular organisms need to carry out to stay alive are performed by different types of cell. This means different cells have different functions.

Most cells do not look exactly like the basic plant or animal cell. They have parts to make them more efficient at their functions. Each type of cell has a special function so we say that the cells are **specialised**. Inside your body there are over 200 different types of cell, all with different functions.

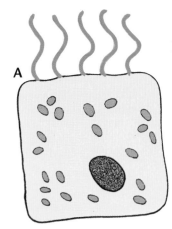

A Some cells have tiny hairs (called **cilia**). These cells are called **ciliated epithelial cells**. Females have cells like these in their egg tubes. The cilia push the eggs along the tubes.

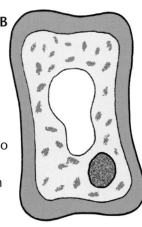

B These cells are packed with chloroplasts containing chlorophyll to trap light energy so that the plant can make food. The cells are tall and thin, so that there are many of them in the leaf. These are called palisade cells.

C For each of cells A and B, decide whether it is an animal cell or a plant cell and give your reasons.

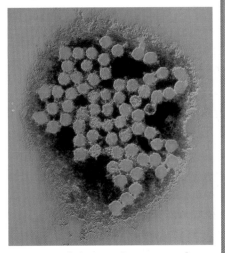
Questions

1 a Make a list of things that are similar in animal and plant cells.

   b Make another list of the differences between them.

2 Which part of a plant cell carries out each function below?

   a controls the cell b gives the cell shape c keeps the cell firm
   d makes food for the plant e controls movement into and out of
   the cell f place where chemical changes happen.

3 a Explain why some cells have:

      (i) chloroplasts **(ii)** cilia.

   b Do all plant cells have chloroplasts? Explain your answer.

4 Viruses cannot reproduce on their own. Explain why this makes some scientists suggest that viruses are not living things.

5 Imagine you are Robert Hooke. Write a letter to a friend explaining why your microscope is so important to scientists.

For your notes:

- All organisms are made of cells.

- There are two types of cell: **animal cells** and **plant cells**.

- Both types of cell have a **cell membrane**, **cytoplasm** and a **nucleus**.

- Plant cells also have a **cell wall** and a **vacuole**. Many plant cells have **chloroplasts**.

- Some cells have special features so that they can carry out their functions.

How living things grow

Living things start small and get bigger. This is called **growth**. You started off life as a single cell the size of the full stop at the end of this sentence. You have grown a lot since then!

This growth is achieved by:

- increasing the number of cells

- increasing the size of cells between divisions.

Increasing the number of cells

If we use a microscope we can observe living cells dividing. This is called **cell division**. Cell division is very easy to observe in organisms with only one cell. Pictures **A** to **E** show one yeast cell making two and then four cells. Photo **X** shows another unicellular organism immediately after cell division.

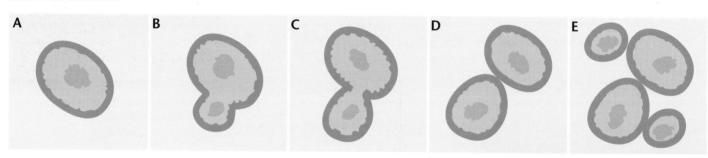

a How many cells would there be if all the yeast cells in E divided again?

b Use the scale bar on photo X to estimate the length and the width of one cell.

It is much more difficult to see cell division in plants and animals, because the cells are grouped together in tissues. Photos **F** to **H** show a plant cell dividing. Photos **I** to **K** on page 7 show an animal cell dividing. The cells have all been stained with dyes to make the parts of the cell show up better.

c It is hard to see the cell membrane in photos F to H. What part of the cell can be seen around the outside of each cell?

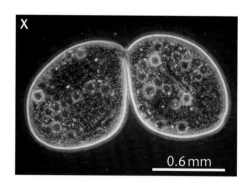

X

0.6 mm

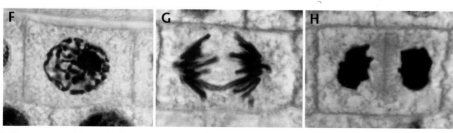

Did you know?

Humans start life as a single cell. An adult human contains about 50 000 000 000 000 cells.
A single cell would have to divide 46 times to make enough cells for an adult human.

Look at photos **I** to **K**. When a cell divides, the nucleus divides first. This is because the nucleus controls the cell. It contains all the information to make sure the cell runs properly. It is very important that every cell has a complete nucleus, with a full set of instructions. During cell division the cell makes two nuclei, and then the cytoplasm and the membrane divide so that one nucleus is in each new cell.

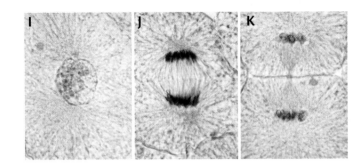

Cells increase in size

When a cell divides, it makes two smaller cells. If you compare photos **F** and **H** you can see that the two new cells (**H**) are smaller than the original cell (**F**). It is the same if you compare **I** and **K**. However, the cells do not stay smaller. They increase in size until they are big enough to divide again.

This is shown clearly with the yeast cells. Yeast cells do not divide into two equal parts. Instead the new cell starts off smaller. Look at photos **B** to **D**. The new cell starts as a little bud and then increases in size.

d What would happen if a cell kept dividing but did not increase in size between divisions?

Questions

1 a Look pictures **M** to **O**. Write the letters in the correct order to show cell division.

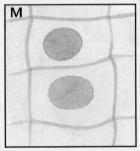 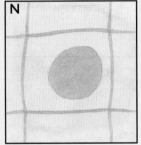 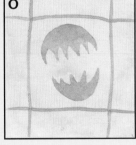

b Do **M**, **N** and **O** show plant cells or animal cells? Explain your answer.

c Which part of the cell divides first?

d Why does this part of the cell divide first?

e The cells look purple because they have been stained with a dye. Why was this done?

f Multicellular organisms grow using cell division, as shown in pictures **M** to **O**. What else happens to cells when organisms grow?

2 a This series of numbers shows cell division.

   1 2 4 8 16 ...

   What are the three next numbers in the series?

b A bacterium divides once every 20 minutes. Starting with one bacterium, how many bacteria will there be after two hours?

Did you know?

Many scientific words are Greek or Latin. This means they use Greek and Latin ways of showing plurals.

Singular	Plural
nucleus	nuclei
cactus	cacti
fungus	fungi
bacterium	bacteria
alga	algae

For your notes:

- All cells are made from other cells.

- Cells divide into two to make more cells. This is called **cell division**. The nucleus divides first.

- Cells increase in size after they divide.

- **Growth** is a combination of increasing the number of cells and increasing the size of the cells.

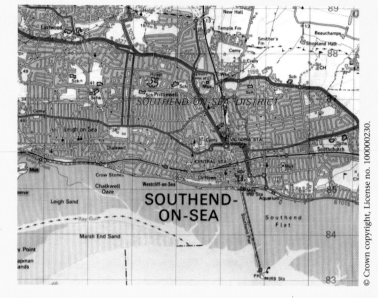

Too small to see

A cell is smaller than the end of a pencil, so it is impossible to draw a cell at its real size. We need to draw it much larger than it really is. Scientists call this a **scale diagram**.

Scale diagrams are also useful for showing things that are too big to fit on a page. We draw big things smaller than they really are. Maps are scale diagrams.

a Discuss in your group other uses for scale diagrams.

Scales

Scaling up means showing an object bigger than it really is. A microscope scales things up. **Scaling down** means showing an object smaller than it really is.

The picture shows Mr and Mrs Beetroot with their two children Nick and Aileen. They are not really this small. They have been scaled down. To find out how big they really are, we have to scale up again.

In this picture, 1 cm is used to show 40 cm in real life. In the picture Mr Beetroot is 5 cm tall. To find his real height you multiply his height in the picture by 40. This means you scale up by 40.

Instead of using lots of words to describe scaling up, we can say the **scale factor** for the picture is 40. This means you multiply by 40. So 1 cm represents 40 cm.

Mr Beetroot

Mrs Beetroot

Nick

Aileen

> 5 × 40 = 200 So Mr Beetroot is 200 cm tall.

b Copy and complete the table to find the real heights of the rest of the Beetroot family.

When we want to scale up, we multiply by a scale factor. Now imagine we want to scale down.

Name	Picture height in cm	Scale factor	Picture height x scale factor	Real height in cm
Mr Beetroot	5	40	5 × 40	200
Mrs Beetroot	4	40		
Nick	3	40		
Aileen	2	40		

c Discuss in your group how you think we can scale down using the scale factor.

If you want to draw Mr Beetroot scaled down by a scale factor of 20, you divide his real height by 20.

$200 \div 20 = 10\,cm$ So you would draw Mr Beetroot 10 cm tall.

d Draw a table with these headings and work out how big you would draw the rest of the Beetroot family.

Name	Real height in cm	Scale factor	Real height ÷ scale factor	Picture height in cm
Mr Beetroot	200	20	200 ÷ 20	10

Finding the scale factor

If you know the real size and the picture size of something, then you can find the scale factor using the formula:

$$\text{scale factor} = \frac{\text{real size}}{\text{picture size}}$$

e Nick has to find the scale factor for a number of household objects. Copy and complete Nick's table shown below.

Object	Real height in cm	Picture height in cm	Real height ÷ picture height	Scale factor
science book	30	3		
house	800	40		
TV	60	12		

Questions

1 Aileen drew her doll 5 cm high. Its real height is 50 cm. Find the scale factor she used to draw the picture.

2 The following table lists some of the objects in the Beetroots' house.
Copy and complete the table.

3 Measure your height in centimetres. Work out how you could draw a scale diagram to represent your height. Draw a line to show your height scaled down by a factor of 10.

Object	Real measurement in cm	Picture measurement in cm	Scale factor
length of car	300	10	
length of pencil		2	10
width of garden	600		50

Parts of a flower

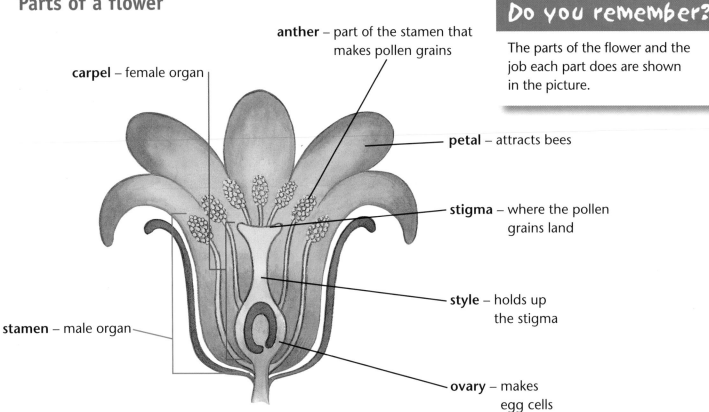

anther – part of the stamen that makes pollen grains

carpel – female organ

petal – attracts bees

stigma – where the pollen grains land

style – holds up the stigma

stamen – male organ

ovary – makes egg cells

Pollination

The job of a flower is **reproduction**, making more plants. The male organs in flowers make **pollen grains**, the male sex cells. You can see pollen grains in the photo on the right. The female organs in flowers make **egg cells**, the female sex cells.

The strongest new plants are made if the pollen grains from one plant join with the egg cells from another plant. To make sure this happens, flowers do not make pollen grains and egg cells at the same time. The pollen grains from an anther on one plant have to be transferred to the stigma on another plant. This is called **pollination**.

a Which insect often helps pollinate flowers?

Fertilisation

Pollination puts the pollen grain on the stigma. The pollen grain then grows a **pollen tube**, as shown in the diagram on the next page. The nucleus from the pollen grain travels down the pollen tube to the ovary.

Fertilisation happens when the nucleus of the male sex cell joins with the nucleus of the female sex cell. This makes a fertilised egg cell.

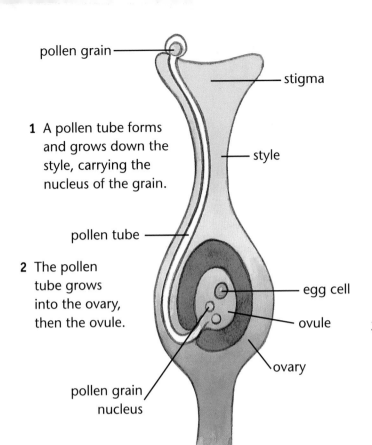

pollen grain

stigma

1 A pollen tube forms and grows down the style, carrying the nucleus of the grain.

style

pollen tube

2 The pollen tube grows into the ovary, then the ovule.

egg cell

ovule

ovary

pollen grain nucleus

The nucleus of the fertilised egg cell contains all the information needed to make a new plant. Half comes from the pollen cell nucleus and half comes from the egg cell nucleus. This information controls what the new plant will look like.

ⓑ Explain the difference between pollination and fertilisation.

3 The nucleus of the pollen grain joins with the nucleus of the egg cell and fertilises it.

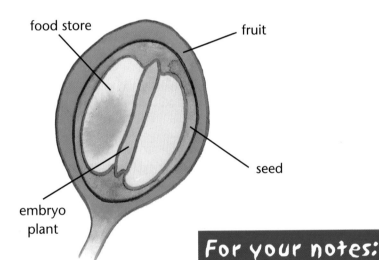

food store

fruit

seed

embryo plant

A new plant

The new fertilised egg cell grows to form an **embryo plant**. The ovule forms a **seed** with the embryo plant inside it. The seed protects the embryo plant and contains a food store for the tiny plant. The ovary forms the **fruit** with the seed inside it.

ⓒ Where will the new plant get its food from once the food store is used up?

For your notes:

- Flowers contain the sex organs of the plant: the male sex cells are the **pollen grains**, the female sex cells are the **egg cells**.

- **Pollination** is the transfer of pollen grains from the **anther** to the **stigma**.

- **Fertilisation** happens when the nucleus of a pollen grain joins with the nucleus of an egg cell.

Questions

1 Draw a flow diagram to show how seeds are formed. Include the following words:

   **stamens   pollen   grain   carpel   egg cell   pollination
   pollen tube   fertilisation   embryo   plant   bee**

2 a What is the function of egg cells and pollen grains?

   b Egg cells and pollen grains have 'half-sized' nuclei with a half set of information. Why is this?

3 Describe how the ovule changes after fertilisation.

4 What are the functions of the seed?

B1 Spot the difference

Why reproduce?

Do you remember?

The wind and insects carry pollen from one flower to another so that the flower can make seeds and the seeds grow into new plants.

Just like plants, animals need to reproduce or they will die out. Giant pandas are becoming very rare in the wild and zoos all over the world are trying to breed them so that they will not become extinct.

More of the same

Most animals need a male and a female to reproduce. Humans have children that grow into adults. The bodies of men and women are different so that that they can produce children. Men have a male reproductive system and women have a female reproductive system.

Men produce sex cells called **sperm**, and women produce sex cells called **eggs**. To make a baby, a sperm and an egg must join together.

a What is the first stage in making a baby?

Male reproductive system

The picture shows the male reproductive system. It is shown from the side.

Sperm are made in the **testes**. There are two of these (one on its own is called a **testis**). The testes produce millions of sperm every day.

The testes are in a special bag of skin called the **scrotum**. This keeps the sperm at just the right temperature. When it is cold, the scrotum gets tighter to keep the testes close to the body and so keep them warm. When it is warm, the scrotum gets looser to move the testes away from the body and so keep cool.

When the sperm leave the testes, they pass along a tube called the **sperm tube**. This carries the sperm to the penis. On the way the sperm pass two **glands**. These glands add a liquid to the sperm. The sperm and liquid together are called **semen**.

Eventually the sperm pass through the **penis**. This is where they leave the man's body.

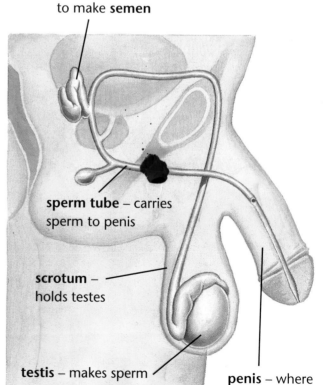

glands – add liquid to make **semen**

sperm tube – carries sperm to penis

scrotum – holds testes

testis – makes sperm

penis – where sperm leave the man's body

Female reproductive system

The picture shows the female reproductive system. It is shown from the front.

The eggs are made in the **ovaries**. There are two ovaries, one on each side. Once a month an egg leaves one of the ovaries and passes along the **oviduct** (egg tube). This takes a few days. The egg and sperm may meet in the oviduct.

The **uterus** (womb) is where the baby develops. The opening of the uterus is called the **cervix**. This is a ring of muscle that can open quite wide to let the baby out through the **vagina** when it is ready to be born.

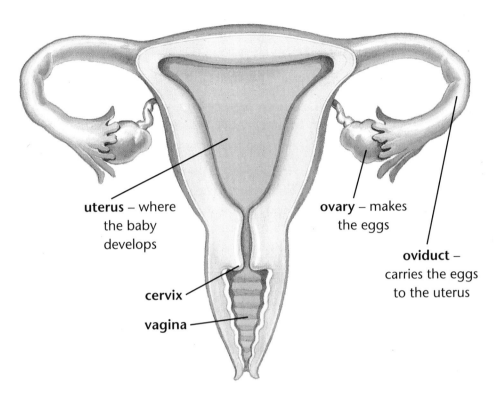

uterus – where the baby develops

ovary – makes the eggs

oviduct – carries the eggs to the uterus

cervix

vagina

Infertility

Some couples cannot have a baby naturally. This is called **infertility**.

Often women have a blocked oviduct that prevents the sperm meeting the egg. This blockage can often be removed by surgery.

But sometimes women may release eggs less frequently than once a month. They can be given drugs that make them release lots of eggs in one go. Then there is more chance of a sperm meeting one of the eggs. If more than one egg is fertilised, the woman can be pregnant with five, six or even seven babies!

Some men have a low sperm count. They produce fewer sperm than other men and so there is less chance of a sperm meeting an egg.

Questions

1 Where are
   a sperm   b eggs
   made?

2 What happens in the uterus? Why is the cervix made of muscle?

3 Give three differences between the male and female reproductive systems.

4 Describe two reasons why a woman might have difficulty becoming pregnant. Suggest what might be done to help her.

5 Explain a frequent cause of male infertility. Suggest a way of overcoming this problem.

For your notes:

● Just like plants, animals need to reproduce or they will die out. To do this they have male and female parts.

● Men make **sperm** in the **testes**. The sperm pass along the **sperm tube** and out of the **penis**.

● Women make **eggs** in the **ovaries**. The eggs pass along the **oviduct** to the **uterus**.

● **Infertility** is when a couple cannot have a baby naturally.

Learn about:
- Sperm and egg cells
- Fertilisation

Sperm and egg cells

The photos show sperm cells and an egg cell seen under a microscope.

Both sperm and egg cells have a nucleus, cell membrane and cytoplasm like any other cell, but they also have special features to help them perform their different functions better. We say that they are **specialised**.

Sperm cells have long tails to help them swim to the egg. They have a small amount of cytoplasm. This reduces their size and gives them a streamlined shape. The pointed shape of the head also helps them to swim to the egg. The head contains chemicals to break down the outer layers of the egg.

Egg cells are round and much bigger than sperm cells. Their larger size allows them to store food in the cytoplasm. They have a protective layer so that only one sperm can get through.

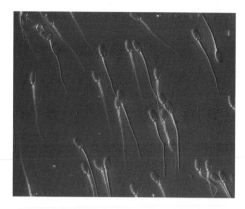

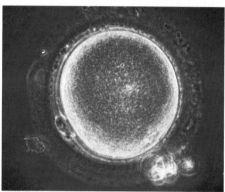

How sperm and egg meet

To make a baby, the male and female sex cells must meet and join together. When a man and a woman make love, the man's penis enters the woman's vagina. Sperm are released from the penis into the vagina. This is how the sperm get into the body of the woman. It is called **sexual intercourse**. The sperm then swim towards the egg.

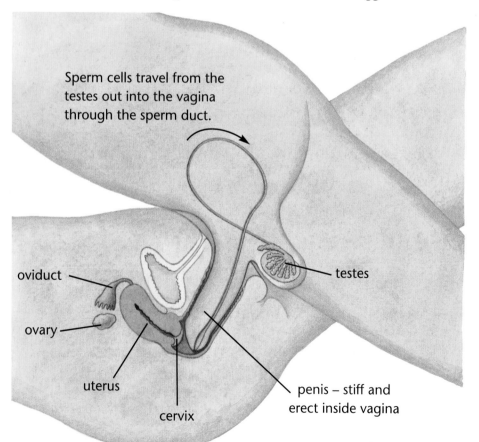

Sperm cells travel from the testes out into the vagina through the sperm duct.

oviduct

ovary

uterus

cervix

testes

penis – stiff and erect inside vagina

Did you know?

During sexual intercourse, up to 500 million sperm are released into the vagina.

What happens next?

The sperm swim up from the vagina into the uterus and then into both oviducts. Many sperm die on the way.

What happens next depends on whether an egg has been released into the oviduct.

If there is an egg in the oviduct the sperm will surround it, as shown in photo A. The first sperm to reach the egg burrows into it. Photo B shows this. The other sperm will die. The nucleus of the sperm fuses with the nucleus of the egg. This is called **fertilisation**. The fertilised egg will become a baby. The woman is pregnant.

If there is no egg in the oviduct all the sperm will die in a short time. No baby will be produced.

2 The egg passes along the oviduct.

3 A sperm meets the egg in the oviduct and fertilises it.

4 The fertilised egg passes along the oviduct to the uterus.

1 An egg is released into the oviduct. Sperm are released into the vagina during sexual intercourse. They swim up through the uterus.

ovary

oviduct

uterus

vagina

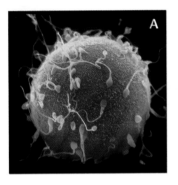

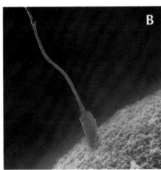

a Explain what fertilisation means.

b What happens to the sperm if there is no egg in the oviduct?

Half and half

The egg and the sperm nuclei each carry half the information needed to make the baby. The sperm contain information from the father and the egg contains information from the mother. So, both parents contribute to the characteristics of the new baby. This is why babies are always similar to their parents. They are not identical to their parents because the information passed on by each sperm and egg is not always the same.

Questions

1 Explain how sperm get into the woman's body.

2 Describe two ways in which sperm cells and egg cells are adapted to their functions.

3 Why do you think so many sperm are produced?

4 Imagine that you are a sperm. Write a story about your journey to the egg. Include all the parts of the body that you swim through.

5 Explain how both parents contribute to the characteristics of a baby.

For your notes:

- Sperm and egg cells are adapted to help them to perform their functions better. They are **specialised**.

- In **sexual intercourse**, millions of sperm are released into the woman's vagina. Most will die, but one may make it through the uterus and to an egg in the oviduct.

- **Fertilisation** happens when the nucleus of a sperm fuses with the nucleus of an egg.

B3 Pregnancy

From egg to baby

After an egg is fertilised, it settles in the thick, soft lining of the uterus. This is called **implantation**.

The fertilised cell divides into 2, then 4, then 8 cells and so on, until there is a tiny ball of cells called the **embryo**. When this happens the woman is **pregnant**. The embryo then grows more to become a **fetus**.

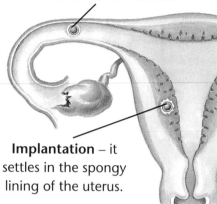

The fertilised egg passes along the oviduct into the uterus.

Implantation – it settles in the spongy lining of the uterus.

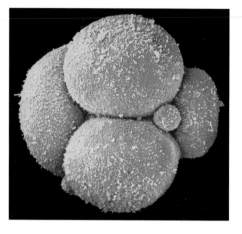
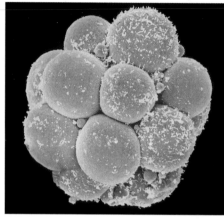

ⓐ What happens to the egg after fertilisation?

The growing baby

The photos below show the development of the fetus during **pregnancy**. (The photos are not to scale.)

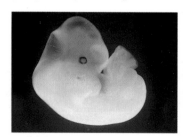

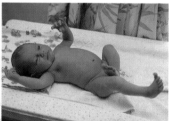

At about 4 weeks the embryo's heart starts to beat. It has eyes, ears and legs.

By about 9 weeks, the baby is called a fetus. It has a head, arms and legs. Fingers and toes start to develop.

At around 22 weeks, the doctor can hear the baby's heartbeat. Its lungs are starting to develop. Its mother will feel it kicking.

At 39 weeks, when it is born, the baby is fully developed. It has a lot of fat to keep it warm when it is born.

ⓑ What differences can you see between the pictures at 4 weeks and 39 weeks?

ⓒ Sometimes babies are born before 39 weeks, perhaps as early as 26 weeks. What special care do you think these babies need?

Getting what it needs

The fetus gets all the substances it needs from the mother's body, through the **placenta**. This forms in the uterus early in pregnancy. The **cord** joins the fetus to the placenta.

The blood in the cord carries food and oxygen to the fetus from the placenta, and it carries carbon dioxide and other waste substances back.

It is important for the mother to look after her health during pregnancy. For example, it is very dangerous for a mother to smoke. Harmful chemicals from the cigarette smoke can cross the placenta and harm the baby.

The diagram on the right shows the fetus just before birth. The fetus usually lies upside down with its arms tucked close to its body. This is the best position for an easy birth.

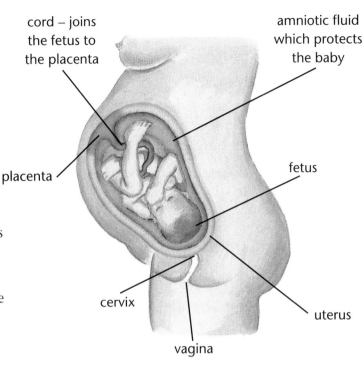

cord – joins the fetus to the placenta

amniotic fluid which protects the baby

placenta

fetus

cervix

uterus

vagina

 d How does the fetus get oxygen and food and remove waste such as carbon dioxide?

Birth

Pregnancy lasts for about nine months. Then the baby is born. It is pushed out of the uterus by **contractions**. These happen when the strong muscles of the uterus wall squeeze. The baby is usually born head-first. After the cord is cut, the newborn baby gets its oxygen from breathing air. The placenta leaves the uterus a few minutes later. This is called the **afterbirth**.

Depending on mother

The mother's body has to adapt ready for when the baby is born. Her breasts grow, preparing for breast feeding after birth. Breast milk is very nutritious. It contains **antibodies**. These are substances that protect the baby from catching common diseases.

For your notes:

- It takes nine months for a human baby to develop fully inside its mother. This is called **pregnancy**. The **fetus** is joined to the mother by the **cord** and the **placenta**.

- At birth, the baby is pushed out of the uterus by strong **contractions**. The placenta then leaves the uterus as **afterbirth**.

- Breast milk contains **antibodies** that protect the baby from catching common diseases.

Questions

1 What is the job of each of these parts?

   a amniotic fluid **b** cord **c** placenta **d** afterbirth.

2 Why does the uterus wall need strong muscles?

3 Describe the ways the mother's body changes during pregnancy.

4 Why do you think that breast feeding is better than bottle feeding?

5 Produce a leaflet explaining to parents how the baby develops inside the mother and how it is born after nine months.

What causes periods?

As we grow up physical changes happen in our bodies. The changes make it possible for us to have babies.

At around 10 to 14 years girls start a monthly cycle called the **menstrual cycle**. The cycle lasts about 28 days and is controlled by substances called **hormones**. Hormones cause an egg to develop and be released in each cycle. The lining of the uterus builds up and becomes soft and spongy.

If the egg is fertilised, it becomes implanted in this lining. If the egg is not fertilised, it dies and the soft lining is not needed for the embryo. The lining of dead cells and blood breaks down and leaves the body through the vagina. This is known as a **period**. We say that the period starts on day 1 of the cycle, as the diagram on the right shows.

a What is a period?

b In a 28-day menstrual cycle, on what day will the egg be released?

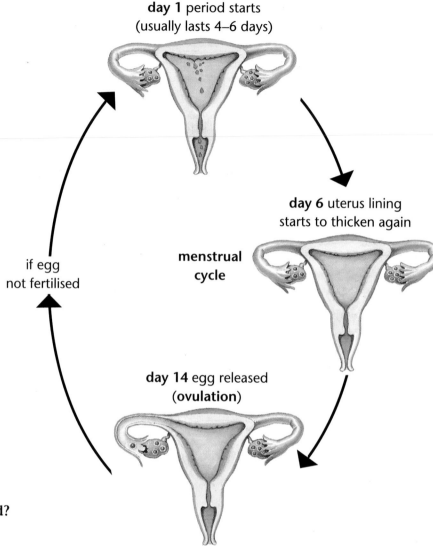

day 1 period starts (usually lasts 4–6 days)

day 6 uterus lining starts to thicken again

menstrual cycle

day 14 egg released (ovulation)

if egg not fertilised

Becoming pregnant

If a woman has sexual intercourse between days 13 and 15, and any sperm reach the oviduct, one sperm will be able to fertilise the egg when it is released. If the egg is fertilised the uterus lining is needed to protect the embryo. It does not break down, so periods stop during pregnancy.

When will it happen?

Kellie started her periods in March during Year 7. She marked the date with a circle in her diary. She marked the first day of her next two periods with a cross.

c (i) How long is Kellie's cycle? (ii) On what date do you think Kellie's fourth period will start? (iii) On what date do you think her first egg might have been released?

One at a time

Usually, only one egg is released and fertilised at a time. This is because the human reproductive system is designed to make one baby at a time. A single baby has a better chance of survival.

Twins

Sometimes a woman gives birth to more than one baby at the same time. Two babies together are called **twins**.

Identical twins, like Hannah and Mary, are produced from just one egg. The egg splits into two just after fertilisation. Because both Hannah and Mary came from the same egg and sperm, they look exactly the same.

Non-identical twins, like Charlie and Amy, are produced if two eggs are released at the same time. Each egg is then fertilised by a different sperm. These twins are no more alike or different than any other brothers and sisters.

The menopause

Women usually stop having periods between 45 and 55 years. This is called the **menopause**. They no longer produce eggs and so cannot become pregnant, so there is no need for the uterus lining to thicken and break down. The menopause is the body's way of adapting to the fact that an older woman is more likely to find pregnancy and looking after a child more tiring than a younger woman.

Questions

1 Explain what happens in the menstrual cycle:

   a between day 1 and day 4

   b at about day 14

   c at about day 6.

2 The diagram represents a 26-day menstrual cycle.

   ● Copy the diagram.

   ● Colour in red the days of the period.

   ● Predict when ovulation will take place and circle the day.

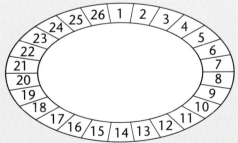

3 In which part of the menstrual cycle is a women most likely to become pregnant? Explain your answer.

4 Suggest reasons why a woman's periods stop when she is pregnant.

5 Explain the difference in the way identical twins and non-identical twins are produced.

For your notes:

● The **menstrual cycle** is a monthly cycle in women, controlled by **hormones**. During the cycle an egg is released, and the woman has a **period**.

● **Identical twins** come from one egg that splits into two after fertilisation.

● **Non-identical twins** are made when each egg is fertilised by a different sperm.

● The **menopause** is when a woman stops producing eggs and her periods stop. She can no longer become pregnant.

19

B5 Adolescence

Learn about:
- Adolescence
- Puberty
- Growth

Take good care

Human babies need a lot more care from their parents than other newborn mammals. They cannot feed themselves or walk when they are born. But baby zebras and giraffes can run around soon after they are born. Unlike humans, they need to do this to avoid being eaten by other animals. Humans look after their young until they are aged 18, so they have a good chance of surviving!

Do you remember?

Growth and development of humans happens in stages as we grow older. First babyhood, then childhood, followed by adolescence and finally adulthood.

All change!

You are all at the **adolescence** stage. Adolescence is a time in everyone's life when physical and emotional changes happen. The changes prepare us to be young adults. The changes happen at different times in different people.

Adolescence finishes when people stop growing, at about the age of 18 years.

a What happens in adolescence?

During puberty

Puberty is the first part of adolescence. It is a time of great change for boys and girls. It is when most of the physical changes happen that make it possible to have babies. Substances called hormones bring about these changes. The testes in boys make the hormone **testosterone**. The ovaries in girls make the hormone **oestrogen**.

b What are hormones?

Puberty usually starts earlier in girls than it does in boys. In puberty, young people often find that their emotions and behaviour change. They become more attracted to the opposite sex.

The changes that happen at puberty are listed in the table. Some are also shown in the pictures below.

Changes in boys	Changes in girls
sudden increase in height (growth spurt)	sudden increase in height (growth spurt)
hair starts to grow on body, including pubic hair	hair starts to grow on body, including pubic hair
voice deepens	breasts grow
testes start to make sperm and hormones	ovaries start to release eggs and make hormones
shoulders broaden	hips widen
sexual organs get bigger	periods start

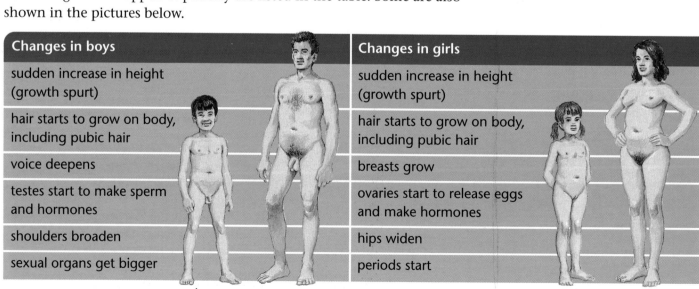

boy → man   girl → woman

Size matters

When a person grows, their cells split into two. This continues to produce more cells. The process is called cell division. At first there is no increase in mass because a big cell divides into lots of smaller new cells. Then the cells increase in size, so the mass increases. This is **growth**.

Growth spurts

A growth spurt is a time of rapid growth. It happens when cells divide rapidly and get bigger. The fetus has a growth spurt. Adolescence is another time when we have a growth spurt.

We can investigate growth by measuring the increase in height or mass of someone. A graph of human growth based on increase in mass would look like this one:

C (i) **How many growth spurts are there?**

(ii) **At what ages do they happen?**

(iii) **When do we stop growing?**

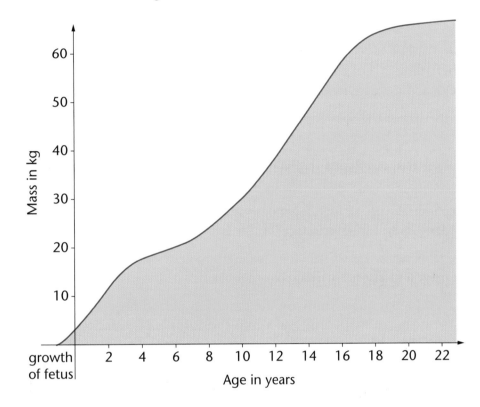

Questions

1 a What two types of changes happen in adolescence?

b What is the first part of adolescence called?

2 a Describe three changes that happen to boys during puberty.

b Describe three changes that happen to girls during puberty.

3 What are the hormones called that cause the changes in puberty in boys and girls?

4 The different parts of the human body do not all grow at the same rate from babyhood to adulthood. Which part do you think is bigger during babyhood. Suggest a reason for this.

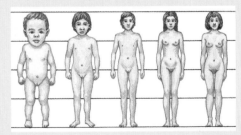

5 Think about each of the following stages of the human life cycle:

babyhood   childhood   adolescence   puberty   adulthood

Draw a table to summarise the physical changes that take place at each stage.

For your notes:

● **Adolescence** is a time when physical and emotional changes happen.

● **Puberty** is the first part of adolescence when most of the physical changes happen.

● The testes in boys make the hormone **testosterone**. The ovaries in girls make the hormone **oestrogen**.

● **Growth** happens when cells divide and increase in size.

B6 Pregnant pause

Time to develop

The length of time that an animal is pregnant is called its **gestation period**. It is the time from fertilisation to birth.

A mouse is pregnant for only three weeks. An elephant is pregnant for nearly two years!

Think about some reasons for a longer gestation period:

● the size of the baby animal. Larger offspring are made up of more cells and have grown a lot before birth.

● how well developed the baby is. There may be a high level of cell specialisation, which takes time.

● parental care. How independent is the newborn animal and what dangers will it have to face?

The table below shows data for the length of time that different animals are pregnant. The animals are in order of size.

ⓐ Plot a bar chart showing the gestation period for each animal. Put the animal along the x-axis (bottom) and gestation period on the y-axis (side).

Now look carefully at the pattern of the results.

ⓑ Describe the relationship between animal size and gestation period.

ⓒ Explain why you think this happens, using as much science as you can.

ⓓ Are there any results that look out of place?

Did you know?

Kangaroos have a very short gestation period. They give birth after about five weeks, but the baby is still much too helpless to survive on its own. It crawls into a pouch and stays in there for another six months. The koala and wallaby also do this.

Did you know?

Rabbits have a shorter gestation period than humans. They can also have more babies at once. Rabbits have a double uterus, and each side can have several baby rabbits developing in it at the same time.

Animal	Gestation period in days
mouse	21
squirrel	30
cat	62
kangaroo	40
ape	200
human	280
camel	355
rhino	420
elephant	649

ⓔ What do you already know about this animal that may explain why it doesn't fit the pattern?

ⓕ Can you predict the gestation period of a rabbit?

If you are not sure about the size of a rabbit it is difficult to predict its gestation period exactly using the bar chart.

Another way of analysing data is to plot a line graph. It is easier to predict using a line graph than a bar chart, but we need different data.

Predators and prey

Here is some data for three more animals. They are all **predators**. This means that they hunt and eat other animals – their **prey**.

Animal	Average adult mass in kg	Gestation period in days
cheetah	95	60
lion	190	108
tiger	210	109

The line graph of this data on the right has adult mass along the x-axis and gestation period along the y-axis. You can predict the gestation period for a different animal, if you know its average adult mass.

The table below shows data for three prey animals.

g Copy the graph and plot the data from the table on it. Join them with a curved line. Label this line 'prey'.

h Do your graphs show a pattern? If so, describe it.

i Which animals tend to have the longer gestation period compared with their body mass, predators or prey?

j An impala has an average adult mass of 55 kg. What do you think its gestation period will be (an impala is a prey animal)?

k Why do you think a line graph is useful?

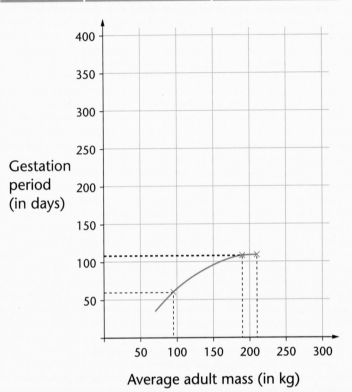

Gestation period (in days)

Average adult mass (in kg)

Animal	Average adult mass in kg	Gestation period in days
antelope	45	180
wildebeest	200	255
zebra	280	360

When lions and tigers are born, they are blind and helpless for a week or so. They cannot move very far, so the mother has to stay in one place to look after them, or carry them around. A baby zebra is very different. It can walk with the rest of the herd only a few hours after it is born.

l Prey animals are at risk from predators. Prey animals are pregnant longer than predators. Suggest two ways this helps the prey animals.

C1 Environments

Learn about:
- Environment
- Adaptation
- Habitat

The environment

The **environment** is the world around us. It includes everything, living and non-living. Sunlight, water, soil, rock, air and all living things are included in the environment.

- A desert is an environment that gets less than 25 cm of rainfall per year. There are hot deserts, like the Sahara, and cold deserts, like central Antarctica.

- A tropical rain forest receives more that 127 cm of rainfall per year. To be 'tropical' the forest must also be close to the **Equator**. This means that is hot as well as wet.

- Other environments include ponds, the sea, beaches, woodlands and grasslands.

a How are the rainfall and temperature different in central Antartica and in a tropical rain forest?

Living in the desert

Many organisms live in desert habitats, like the North American desert shown in the photo below. These living things have special features that suit them for living in the desert. We say that they are **adapted** for living in the desert. Their special features are **adaptations**.

The saguaro cactus in the photo on the right has adaptations. The roots stay at the surface, ready to take in water when it rains. They spread outwards to 30 metres in all directions. The saguaro cactus can store a lot of water. Like many cacti, the saguaro does not have leaves. This is because plants lose a lot of water through their leaves by evaporation.

b Explain how a saguaro cactus survives in the desert.

Desert animals have to avoid over-heating in the high daytime temperatures (up to 75 °C) as well as dying from lack of water. Look at the kangeroo rat photo. They lose very little water in their sweat and urine. They dig burrows that are 30 cm deep into the cooler, moister sand and plug the entrances with bits of cactus to keep up the humidity.

Do you remember?

An **organism** is a living thing, such as a plant or animal.
A **habitat** is where an organism lives. A **condition** is a variable in a habitat that can be measured – for example, temperature, light level and rainfall.

Kangaroo rats are **nocturnal** – they come out only at night, after the Sun has set. They dry the seeds they collect in the hot, surface sand before storing them in their burrows. The dry seeds absorb water from the surrounding sand, increasing the water in the kangaroo rat's diet.

c Choose three of the kangaroo rat's adaptations and explain how they help it to survive in the desert.

Jerboas live in deserts in Africa, Asia and South America.

Gerbils live in African deserts.

Jerboas and gerbils are other rodents that live in deserts. They look similar and behave in a similar way because it is the best way for rodents to survive in a desert.

For example they all have wide, hairy feet for a good grip on the sand, and strong back legs so that they can jump long distances. This combination of grip and push means that they can move fast and change direction very quickly.

d Why do you think are these animals adapted to move fast and change direction quickly?

Different environments, different adaptations

Look at the photos on the right. One shows a fox from the North American desert. The other shows an arctic fox from Canada.

e Explain the following differences between the two types of foxes:
(i) colour   (ii) thickness of coat.

Desert fox.

Arctic fox.

Questions

1 Other desert plants have other adaptations. Explain how the following adaptations will help each plant survive.

  a The mesquite bush has roots that are 30 metres deep.

  b The ocotillo bush sheds all its leaves in a drought and grows new leaves only when the rain comes.

  c Some evening primroses live their whole life cycle within a few weeks, when it rains. They germinate, grow and make more seeds during this short time.

2 Insects called desert crickets live in the desert. They …

  ● have very wide feet   ● stay in the shade at midday

  ● are pale brown.

  a Explain how each of these adaptations helps the cricket survive.

  b What other desert animals have wide feet?

  c Crickets in other habitats are often green. Explain this colour difference.

For your notes:

● An organism's environment is everything that surrounds it, including its **habitat**.

● Living things are **adapted** to the **conditions** in their habitat. This helps them survive.

Day and night

Living things have to survive day and night every 24 hours. Some animals are **diurnal**. They are active during the day and rest during the night. Other animals are nocturnal. They are active during the night and rest during the day.

a **Use your general knowledge to name three nocturnal animals that live in Britain.**

In desert environments, there are big differences between the day and the night. Most desert animals are nocturnal because the night is much cooler. The deer horn cactus in the photo flowers only at night. Each flower lasts just one night. The flowers are pollinated by nocturnal insects and bats.

Tides

Another daily change is the tide. There are two high tides and two low tides each day. The beach exposed at low tide is called the **intertidal area**. Living things living in the intertidal area have to be able to cope with large changes in conditions.

When the tide is in ...	When the tide is changing ...	When the tide is out ...
● living things are surrounded by water ● temperatures are constant ● the level of saltiness is constant.	● living things are pushed and pulled by strong currents.	● they are exposed to air and may dry out ● temperatures change ● there is increasing saltiness as the water evaporates.

Martin is interested in the seaweed. It is gooey and slimy. It is stuck to the rocks. It will not come away, no matter how hard Martin pulls. It is covered in blobs that pop when you squeeze them, like bubble wrap. His dad tells him to stop popping the blobs because the seaweed needs them to stay alive.

This photo shows this seaweed, called bladderwrack. The blobs are bladders. They are filled with air and act as floats when the seaweed is underwater.

Bladderwrack is adapted to the daily changes that happen on the beach.

When the tide is in...	When the tide is changing	When the tide is out...
• the bladders float. This keeps parts of the seaweed close to the surface • there is lots of light at the surface, so the seaweed can make its own food.	• the seaweed is stuck to the rock by a holdfast. It cannot be carried away by the current.	• the seaweed is covered with a gooey slime. This stops it drying out.

b How is bladderwrack adapted so it can:
(i) feed? (ii) stay in one place? (iii) not dry out at low tide?

Chloe is more interested in the barnacles. She finds it difficult to believe that they are alive. Look at the photo. They look like part of the rock. Chloe's dad tells her that they are animals. When they get home, they look at pictures of barnacles in a book. Chloe is surprised to see that the barnacles look very different when they are underwater. Inside each bony cone there is a little animal.

The barnacles are adapted to the daily changes that happen on the beach.

When the tide is in...	When the tide is changing	When the tide is out...
• the barnacle waves its feathery feet, collecting food.	• the barnacle is stuck to the rock by its back.	• it closes its bony plates to trap water inside. This stops it drying out.

c How are barnacles adapted to living in the intertidal zone?

Questions

1 Use pages 24 to 27 to answer these questions.

a Compare conditions in a desert during the day and during the night.

b Why does a kangaroo rat stay in its burrow during the day?

c What would happen to the deer horn cactus if it flowered during the day?

2 Rock pools contain seawater that has been trapped between rocks when the tide goes out. Imagine that the tide goes out in the morning, and does not come back in until the evening. How will the conditions in the rock pool change during the day?

3 When covered with water, limpets move about rocks feeding on algae. When exposed to air they cling to the rock, trapping water inside their shell. They even have a 'home' where the rock has been worn away to fit the shell exactly. How are limpets adapted to living in the intertidal zone?

For your notes:

● Conditions in a habitat change over every 24 hours.

● Living things are adapted to the daily changes in their habitat.

C3 Changing seasons

Winter and summer

During winter, the Sun is lower in the sky. The sunlight is less bright and less hot. The days are shorter. This means that it is colder in winter than in summer. Sometimes it is cold enough for snow. Plants and animals need to be adapted to survive the winter.

Do you remember?

Green plants need sunlight and water to make their food.

Off season for plants

In winter there is much less sunlight. Plants can make very little food in the winter. So many plants adapt by shutting down in the winter. They become inactive or **dormant**.

Some plants lose their leaves. We call these plants **deciduous**. Look at the photos of the oak wood on the right. In the winter the plants have lost their leaves.

Some plants survive the winter underground. Daffodils survive underground as bulbs.

Other plants survive the winter as seeds. Poppies survive the winter this way. The seeds grow into new poppy plants during the next spring.

a Why do many plants become dormant in winter?

Toughing it out

Winter is a difficult time for animals. It is cold and food can be hard to find. Many plants have lost their leaves or stopped growing. This means that there is less food for animals that eat plants.

Many animals eat more food than they need during the summer and autumn. They store the extra food as fat and use their fat reserves when food becomes scarce. They also grow a thicker coat of fur for the winter.

Some animals, like squirrels, make a store of food for the winter. They collect food in the autumn, when there are lots of nuts and berries. They then live on this food during the winter.

Some animals hibernate. **Hibernation** is a deep sleep. The dormouse in this photo hibernates in the winter so it does not have to find food.

b How do you think the dormouse prepares for hibernation?

Taking a winter break

Other animals leave winter behind. They move to another habitat with better conditions. We call this **migration**. Many birds migrate, including swallows.

In Britain, birds fly south to avoid winter. In other parts of the world animal migrate because of lack of water rather than cold weather. In Africa, herds of wildebeest migrate every year so they do not run out of water.

Many animals living in the sea also have annual patterns of migration. These include fish like salmon and many types of whale.

Winter white-out

The natural landscape in Britain is a mixture of browns and greens. Many animals blend into this landscape. We say they are **camouflaged**. This helps animals like rabbits hide from predators. It helps animals like foxes creep up on their prey.

Snow turns the brown landscape white. The camouflage doesn't work any more. Prey animals are more likely to be eaten. Predators find it harder to sneak up on their prey.

Some animals change their coats during the winter. Their winter coats are white and much thicker. Look at the photos on the right. They show an arctic hare in summer and in winter. The white coat helps the hare stay active in winter.

Did you know?

The same fin whale was found in February in the Antarctic and in July off South Africa, a journey of over 3000 kilometres!

Do you remember?

A **predator** is an animal that hunts and feeds on other animals. A **prey** animal is hunted by a predator for food.

Questions

1 Why is there less food in winter for:

   a plant eaters?   b meat eaters?

2 a Describe the adaptations that help the arctic hare to survive the winter.

   b Explain how each of these adaptations increases the arctic hare's chances of survival.

   c Explain why arctic foxes also grow white coats in winter.

3 Read the section about migration.

   a Explain why animals migrate.

   b Why doesn't Britain have migratory land animals like wildebeest?

   c Read the 'Did you know?' Explain why the whale is found further north in July and further south in February.

For your notes:

- Conditions in a habitat change with the seasons.

- Some living things live through winter by **hibernating**, **migrating** or becoming **dormant**.

- Other living things have adaptations that allow them to stay active throughout winter.

C4 Adapted to feed

Who eats what?

Living things are either **producers** or **consumers**. Producers include plants and algae (for example seaweeds). Some consumers are **herbivores**. They eat only producers. Other consumers are **carnivores**. They eat only other consumers. **Omnivores** are animals that eat both producers and consumers.

Do you remember?

A producer is a living thing that makes its own food using sunlight. A consumer is a living thing that feeds on other living things.

(a) Name one producer, one herbivore, one carnivore and one omnivore.

Producer adaptations

Plant leaves are broad and thin, so the sunlight can reach all the cells. Leaves are arranged on the stem so that light falls on every leaf. Cells in the leaf contain a green pigment called **chlorophyll** that traps the energy in the sunlight.

Herbivore adaptations

Many herbivores eat leaves and stems. Leaves and stems do not contain a lot of energy, so these animals need to graze for many hours each day. They have teeth that are adapted to eating leaves and stems. Leaves and stems are difficult to digest so many herbivores have microbes living in their guts to help with digestion.

(b) Look at the photos. They show a giraffe and an elephant. How do these animals get leaves and stems that are unavailable to most other herbivores?

pollen basket

pollen comb

Other herbivores live on seeds, fruits, pollen or nectar. These parts of plants contain a lot more energy than the leaves and are a lot easier to digest. The photo on the right shows a bee collecting pollen for food. The bee's hairy body and shaped legs are adapted for pollen collection. The pollen catches on the hairs and is combed off into the pollen baskets.

Carnivore adaptations

Carnivores are predators. They have adaptations that help them find, catch and kill their prey.

- They have acute senses so that they can find their prey. Bats have superb hearing and use echoes (sonar) to locate their prey in the dark. Foxes have excellent smell and hearing.

- Cheetahs can run at over 100 km per hour and foxes are camouflaged so they can sneak up on their prey and catch them.

- Cobras have poison that they inject using their fangs to kill their prey.

Prey adaptations

Prey animals have adaptations to help them escape predators.

- The have acute senses so that they can detect predators approaching. Rabbits have excellent peripheral (all-around) vision and hearing.

- Many have safe places to go. Rabbits live in burrows.

- Many prey animals are camouflaged.

- Many prey animals move very fast and change direction very quickly. Rabbits hop and jump.

- Many prey animals live in groups. The flash of a rabbit's white tail warns other rabbits to run. Stronger members of a wildebeest herd will attack predators, even lions.

C **Wildebeest will trample lion cubs to death if they get the chance. How does this behaviour help the wildebeest survive?**

Questions

1 How do the following adaptations help the producer make its food?

  a Bluebells live in deciduous woodland habitats. They are dormant in winter and grow in very early spring before the trees come into leaf.

  b Ivy climbs up trees as it grows.

2 Look at the photos of the bee on page 30. How do the following features help the bee collect food?

  a hairy body   b pollen comb   c pollen basket.

3 How do the following features help the animal hunt?

  a a spider's web   b a cobra's fangs

  c a bat's sonar   d a cheetah's sprint.

4 Read the information about herbivore, carnivore and prey adaptations on these two pages. Answer the questions, using the evidence.

  a How are rabbits are adapted to eating grass and avoiding predators?

  b How are foxes are adapted to hunting and eating rabbits?

For your notes:

- **Producers** make their own food. **Consumers** eat other living things. Consumers can be **herbivores**, **carnivores** or **omnivores**.

- Living things have adaptations so that they are good at making, getting at or catching their food.

- Predators have adaptations that help them hunt their prey.

- Prey animals have adaptations that help them avoid being eaten by predators.

Food chains

As you know, a **food chain** shows you who eats what. In this food chain the grass is the producer. The rabbit eats the grass and the fox eats the rabbit.

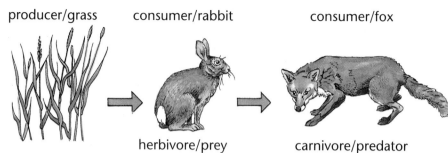

producer/grass   consumer/rabbit   consumer/fox

herbivore/prey   carnivore/predator

Arrows in a food chain show you how the energy is transferred from producer to consumer. Producers get their energy from the Sun. Look at the food chain. Energy from the Sun is stored in the grass.

● When the rabbit eats the grass, some of the energy moves from the grass into the rabbit. The arrow shows this.

● When the fox eats the rabbit, energy moves from the rabbit into the fox. Again the arrow shows this.

So the fox's energy came from the Sun.

Interdependence

This food chain tells us that rabbits are dependent on grass. If the grass died, the rabbits would die of starvation and then the foxes would die.

Look at the food chain. Imagine all the foxes died.

a What would happen to the rabbits in the short term?

b What would happen to the grass?

c What would happen to the rabbits in the long term?

If the rabbits had no predators they would eat all the grass and starve. The foxes would starve without the rabbits, but the rabbit would starve without the foxes. This is one example of **interdependence**.

Food webs

In fact, feeding relationships are more complicated than this. There are many food chains in any one habitat. These food chains overlap. For example, both foxes and grass are involved in many food chains. We represent overlapping food chains using **food webs**. The diagram shows a food web for a woodland habitat.

d How many food chains can you find in this food web?

Food webs show interdependence much more clearly than separate food chains.

Did you know?

Mosses, ferns and seaweeds are producers. Food chains in the sea often start with uni-cellular producers called phytoplankton. All producers contain the green pigment called chlorophyll.

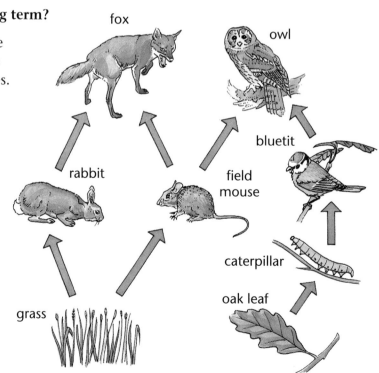

fox   owl

rabbit   field mouse   bluetit

caterpillar

grass   oak leaf

It would be difficult to draw a food web that included all the feeding relationships in a habitat. It would need to show all the producers, herbivores, carnivores and omnivores. It would be huge!

e Look at the woodland food web on the previous page. Where would you add in these living things:
(i) the fleas living in the fox's fur? (ii) bees?

Sorting out feeding relationships

To construct food chains and food webs you have to know what every living thing eats. You can observe living things eating, but others are hard to see because they are very small, or active only at night, or run away from humans.

We can use cameras, including infrared cameras, to observe animals from far away. We can find out where animals go by tracking their footprints or by tagging them with a radio transmitter. It is possible to work out what the animal eats from where it goes. Another important method is to analyse animals' droppings. These often contain bits of food that have not been digested.

f Owls swallow their prey whole and then regurgitate the bits they cannot digest as 'owl pellets'. Suggest four ways that an owl's eating habits could be investigated.

Questions

1 Giant pandas eat only bamboo shoots. What is the danger of an animal having only one food source?

2 **a** Arrange these three living things into a food chain:

- lion
- grass
- antelope.

b Cheetahs eat antelope. Add a cheetah to your food chain to make a food web.

c Elephants eat grass. Add an elephant to your food web.

d Elephants also eat acacia trees. Add acacia trees to your food web.

e Giraffes eat acacia trees, and lions eat young giraffes but not elephants. Complete your food web.

3 Look at the food web for the woodland habitat. Imagine a disease killed all the oak trees.

a (i) What would happen to the caterpillars?
(ii) What would the bluetits eat?
(iii) Would the grass grow better?
(iv) What would happen to the number of fieldmice?

b Robert thinks that owls would have more food to give their young. Carla thinks that owls would have less food to give their young. Explain how Robert and Carla came to their different conclusions.

For your notes:

- **Food chains** show feeding relationships. They also show how energy is transferred from producer to consumers, and from consumer to consumer.

- The energy in a food chain comes from the Sun. The producer uses sunlight to make its food.

- **Food webs** show all the food chains in the same habitat.

- All living things in a habitat are **interdependent**.

Putting into groups

You can look for meaningful patterns by putting things into groups.
If the rules for deciding the groups are good, all the things in that group
will have many features in common.

Sabrina and Nel are trying to group
the living things on the opposite
page. They start by grouping the
organisms by colour. They do this
because Nel suggests that producers
are green.

Green	Red	Brown	Grey
gut weed	dahlia anemone	beefsteak fungus	elephant
bush cricket	coral weed	acorn barnacle	
oak		bladderwrack	

a Are all the living things in the 'green' column producers?

b Are all the producers under 'green'?

c Look at the four groups. Do you think they are useful?
Give your reasons.

Feeding relationships

Nel decides that it is pointless trying to group living things by their
looks. She says that you cannot make a start on grouping living things if
you do not know how they feed.

d Divide the living things into producers and non-producers.

e Divide the non-producers into herbivores, omnivores
and carnivores.

f Did you have problems grouping some of the living things?
Which ones? Explain why.

g Is this grouping useful? Explain your answer carefully.

Other groupings

Nel and Sabrina decide to investigate grouping the living things based
on movement.

h (i) Divide the living things into two groups 'moving' and
'not-moving'. Make a table showing your groupings.
(ii) Is this grouping useful? Do the living things in each group
share a lot of features?

Nel suggests dividing the groups again, this time based on colour and
movement. She suggests four groups: 'moving and green', 'moving and
not-green', 'not-moving and green', 'not-moving and not-green'.

i (i) Produce a table showing Nel's suggestion.
(ii) Are Nel's groupings useful?

Dahlia anemone.
About 5 cm high. Lives stuck to a rock. 80 short tentacles. Hunts other animals for food.

Beefsteak fungus.
7–20 cm wide, 2–5 cm thick. Lives stuck to a tree. Feeds on dead plant material.

Gut weed.
Thin (about 5 mm wide), green tubes. Lives stuck to a rock. Traps sunlight to make food.

Acorn barnacle.
5–15 mm in diameter. Lives stuck to a rock. Combs food particles from the water.

Bladderwrack.
Long, flat, wide (2–5 cm) brown fronds with bladders. Lives stuck to a rock. Traps sunlight to make food.

Oak.
Up to 45 m tall. Green leaves in spring and summer. Lives stuck in the ground. Traps sunlight to make food.

Great green bush cricket.
4–5 cm long. Hops and flies. Feeds on plants and other insects.

African elephant.
6–7 m long (including trunk), 3–4 m tall (to shoulder). Moves about. Feeds on grass and other plants.

Coral weed.
Flat, thin, long (6–7 cm) red fronds. Lives stuck to a rock. Traps sunlight to make food.

Questions

1 Sabrina does some research. She finds out that sea anemones and barnacles move when they are young.

  a Redo the table you made of Nel's groupings of colour and movement, taking this into account.

  b Does Sabrina's extra information help? Give reasons for your answer.

2 Would you put the beefsteak fungus in with a 'producer' group or a 'consumer' group? Give reasons for your answer.

3 Do you agree or disagree with these statements? Give your reasons.

  a 'Green' means producer.

  b 'Green and not-moving' means producer.

  c All producers are green and do not move.

All the same?

Variation and classification

Species

There are millions of living things in the world. To make sense of them all, we have to find a way of putting them into groups. Some living things are very similar in the special parts they have or the things they do. These are called their **features**. When living things have a lot of features the same they are in a group called a **species**.

Members of the same species can mate and **reproduce**. Their offspring will also be able to reproduce. Look at the pictures. Wolves and reindeer are both mammals, but different species. They cannot mate to produce a baby 'wolfdeer' or 'reinolf'!

Did you know?

A horse and a donkey are so similar that they can mate together. The offspring is not a horse or a donkey but a mule. Mules cannot have babies. This tells us that horses and donkeys are different species.

Wolf.

Reindeer.

How do we describe a species?

The features of a particular species may be described in different ways.

A novel might describe an animal like this:

> Reynard was handsome, proud and independent. He led a quiet life, often happy to laze away the day at Sleepy Hollow with his mate Rena and the cubs.
> In the autumn, every evening as night was drawing in, he ventured out for his stroll. His elegant russet coat and bold brush blended with the bracken fern and had a particular way of catching the last of the fading light.
> That night things went a little differently from normal. He pricked his pointed ears but the sound was not the welcome thud of rabbits' feet…

A field study guide for a biologist might describe the same animal in a much more factual way like this:

- red-brown coat
- white chest
- bushy tail
- pointed ears
- eats rabbits
- nocturnal
- solitary
- lives in underground dens

The scientist describes the coat as 'red-brown'. The novelist describes the coat as 'russet'. Another writer might choose the word 'rusty'.

Scientists need to share a system of describing features so that it is clear which species they are talking about. This is why all scientists use the same words, for example 'nocturnal' and 'tail'.

a Which animal do you think the novel and study guide are describing?

b Read the extract from Reynard's story. Identify the text that tells us that Reynard: (i) is solitary (ii) is nocturnal (iii) has a red-brown coat.

c Why is the scientific description more useful if we want to study differences between species?

The same but different

Humans all belong to the same species, *Homo sapiens*, because we have many of the same features. But there are differences between us. We have different colour eyes and hair. We have different weights and heights. Some of us are better at maths than others. Some of us are better at sport.

Differences like these are all examples of **variation** within our species. Some of these differences are passed on or **inherited** from our parents. Others are caused by our surroundings.

d In what ways are the people in this crowd the same?

But no two people are exactly alike – not even identical twins!

e In what ways are the people in this crowd different?

Questions

1 Wolves and reindeer are different species. What differences are there between them?

2 Why do you think that you cannot mate a polar bear with a seal, but you can mate a polar bear with a brown bear?

3 How do we know that horses and donkeys are two separate species?

4 Humans are all the same species. What causes so many differences between us?

5 List three features of humans you think are inherited.

6 Describe the features of your family to set the scene for them to be the main characters in a novel. Emphasise their similarities.

For your notes:

- If there are enough similar **features** between living things, they belong to the same **species**.

- Members of the same species can **reproduce** and the species will continue.

- We call the differences between living things **variation**.

- Some variations between the members of a species are **inherited** from their parents, and some are caused by their surroundings.

37

D2 Differences count

Keep it in the family

The members of the Jones family are very alike. So are the members of the Smith family. Each person looks like their parents. This is because some features, like natural eye colour, are passed on or inherited from our parents.

Billy Jones

Amy Smith

Instructions for your eye colour, skin colour and whether you are male or female were passed to you in the **sperm** cell from your father and the **egg** cell from your mother. When the sperm and egg fuse, there is mixing up of the instructions. This is why each of us inherits some features from our mother and some from our father.

a **What similarities can you see between members of the Jones family?**

We are all different

But the members of the Jones and Smith families are not identical to other members of their families. We are all different because different features are passed on to us from our parents. This is called **inherited variation**. There are variations between the members of a family because each sperm and each egg has a different set of instructions.

b **What differences can you see between the members of the Smith family?**

c **What do you think might have caused two of these differences?**

We are also different because our surroundings or the **environment** we are brought up in causes variation. This is called **environmental variation**. Some of our features are simply down to choices that have been influenced by our surroundings. Dyeing your hair is an example.

Did you know?

Identical twins both inherit the same information from their parents because they develop from the same sperm and egg. But they sometimes have different features because they have grown up in different environments.

Inherited or environment?

Sometimes it is difficult to tell whether a feature has been inherited or caused by the environment. This is because inherited features sometimes skip one or more generations. It is also difficult to tell what a person would have been like if they had been brought up in a different environment.

Tall children inherit their tallness from their parents. But if they do not eat the right sort of food, they will not grow as tall as they should.

Questions

1 Donna has these features: pale skin, small stature, blue eyes, pierced ears, naturally red hair, a tattoo on her arm, a high IQ.

Copy and complete the table to classify Donna's features.

Inherited	Inherited and environmental	Environmental

2 A group of students measured 50 ivy leaves from the same bush. The widths of the leaves ranged from 8 mm to 42 mm. Suggest reasons for this variation.

3 Two identical twins Rebecca and Rachel were adopted at birth by different families. In their twenties they managed to find each other and meet. Rebecca was taller and skinnier than Rachel. What do you think could have caused this?

4 Rebecca and Rachel also wanted to find their natural parents, in particular their mother because they thought they would be more like her than their father. Do you agree with this idea? Explain your answer.

For your notes:

- Some variations between the members of a species are **inherited** from their parents and some are caused by their **environment**.

- Some features can be affected by both inheritance and the environment.

D3 Sorting living things

Learn about:
- Classification
- Vertebrates
- Invertebrates

Living things

We call living things **organisms**. The smallest living things are called **microorganisms** and you need a microscope to study them.

Classifying organisms

There are lots of different species of organisms so we put them in groups to make them easier to study. We put species that have similar features into the same group. This grouping is called **classification**. The groups and their descriptions can help us to name any organisms that we find.

The table shows how we start to classify living things.

Animals	Plants	Microorganisms	Fungi
● human, horse, spider	● lime tree, primrose	● virus, bacterium	● toadstool, mould
● feed on other animals or plants	● make their own food	● can only be seen with a microscope	● feed on rotting material
● must move around	● green		

Wherever you look, you will find examples of all these groups of organisms. You will find **animals**, **plants**, microorganisms and **fungi** in soil or in a pond. There are many different living things even in very cold places like the Arctic.

Animal X-rays

All animals can be put into two smaller groups, those with a backbone and those without a backbone.

The first scientist to classify animals was a Frenchman called Georges Cuvier who lived in the eighteenth century. Cuvier grouped all the animals with backbones together and called them 'Vertebrata', a Latin word for jointed backbone. In classification today, we call animals with backbones **vertebrates**. We call animals without backbones **invertebrates**.

Here are some X-rays of animals from the Arctic.

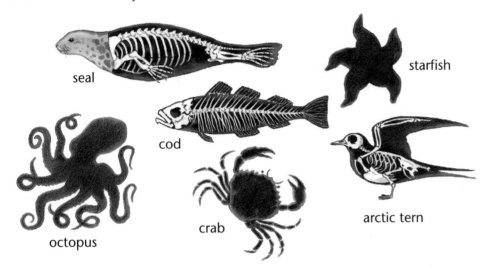

Georges Cuvier.

a Classify each animal as a vertebrate or an invertebrate.

b How do you think Cuvier described an elephant from its tooth? What measurements would he have compared?

Did you know?

Cuvier had a huge collection of animal bones and studied them carefully. He was particularly interested in elephants. He described the exact size and measurements of an extinct type of elephant from one tooth fossil!

Questions

1 Explain how living things are classified.

2 How are plants different from animals?

3 Bacteria are classified as microorganisms, not as plants or animals. Look at this diagram of a bacterium. Suggest a feature that could be used to classify it as:

a an animal   **b** a plant.

cytoplasm   cell wall

nuclear material

no chloroplasts or large vacuoles

4 Fungi look a bit like plants. Why are fungi in a group of their own?

5 Which of these animals are vertebrates?

> earthworm   octopus   robin   toad   rat
> snake   centipede   bat   squirrel

6 Design a key to classify vertebrates and invertebrates.

For your notes:

- We can sort **organisms** into groups with similar features. This is called **classification**.

- **Vertebrates** are animals with a backbone.

- **Invertebrates** are animals without a backbone.

41

D4 More animal groups

Five groups of vertebrates

The vertebrates are divided into five smaller groups:

- ● mammals
- ● birds
- ● reptiles
- ● amphibians
- ● fish

Lion   Eagle   Crocodile   Frog   Shark

Sort yourself out

The human body has a backbone, so we are vertebrates. Along with the lions, apes, dogs, cats and many other furry animals, we are classified as **mammals**.

Mammals

These are the features of mammals:

- ● Their babies develop inside the mother's body.

- ● The mother feeds the young on her milk, which she makes in her **mammary glands**.

- ● They have hairy skin to insulate them. They are warm blooded. This means they keep their body temperature constant at 37 °C.

The rest of the vertebrates apart from mammals are classified as:

- ● birds ● reptiles ● amphibians ● fish.

Did you know?

Mammals and birds are the only groups of vertebrates that look after their young. Reptiles, amphibians and fish usually leave their young to fend for themselves.

Birds

These are the features of **birds**:

- ● They lay eggs with hard shells.
- ● They look after their young after they have hatched.
- ● They have feathers and wings.
- ● Most birds can fly.

Birds and animals are warm-blooded. They have ways of keeping their temperature the same.

Fish

These are the features of **fish**:

- They can only live in water. They lay eggs in water.
- They breathe through gills.
- They have scales and fins.

Amphibians

These are the features of **amphibians**:

- They lay jelly-like eggs in water.
- They breathe air and live partly on land, but have to lay their eggs in water.
- They have a smooth, moist skin.

Reptiles

These are the features of **reptiles**:

- They lay eggs on land. Their eggs have leathery shells.
- They breathe air and live mainly on land.
- They have a scaly, dry skin.
- Reptiles like lizards move around to control their body temperature. Sometimes they sunbathe to warm up, but if it is very hot they shelter from the sun.

Reptiles, fish and amphibians don't create their own body heat. They are the same temperature as the surroundings.

Did you know?

Mammals, birds, reptiles and amphibians breathe air using lungs.

Questions

1 In which vertebrate group are humans? Explain your answer.
2 What features do all birds have in common?
3 Describe a crocodile's skin.
4 Why do salamanders go back to water in the spring? Do you think all amphibians do this? Explain your answer.
5 How does a fish such as a salmon breathe?
6 Why do birds and reptiles have eggs with shells, but humans don't?
7 Give three features that make an amphibian different from a reptile.

For your notes:

- Vertebrates are classified into five groups.
- The groups are **mammals, birds, reptiles, amphibians** and **fish**.
- Each group has different features.

Invertebrates

The invertebrate animals have no backbones. They have no legs or lots of legs. The best way to start to classify them is to look for their legs!

No legs

We can start to sort the invertebrates with no legs into groups by the kind of body they have. Some have hard bodies, others have soft bodies. They also have different shaped bodies. There are six groups:

● **Jellyfish** have a soft jelly-like body.

● **Flatworms** have a soft flat leaf-shaped body.

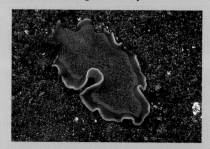

● **Segmented worms** have a soft ringed body.

● **Starfish** have a hard star-shaped body.

● **Roundworms** have a soft, thin round body.

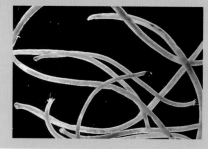

● **Molluscs** have a soft muscular body with one foot. Most have a hard shell.

Jointed legs

We call the invertebrates with lots of jointed legs **arthropods**. Arthropods have bodies made of sections called **segments**. They have a hard outer covering called an **exoskeleton** that sometimes forms a shell. We divide the arthropods into four smaller groups:

● **Crustaceans** have a soft body, usually with a hard shell. They have lots of legs.

● **Centipedes** and **millipedes** have a long thin body. They have lots of legs.

● **Spiders** have a 2-part body and 8 legs.

● **Insects** have a 3-part body and 6 legs.

Cuvier collection

Cuvier put all the sea shells together and called these 'Mollusca'. He called all the insects, spiders and crabs 'Articulata' (Latin for 'jointed'). He called all the other small creatures 'Radiata'.

Where does it fit?

Trilobites are animals that existed before the dinosaurs. They are arthropods with hard shells, a 3-part segmented body and lots of legs. Their bodies have been found preserved in rocks, but many of these are only empty shells. Only a few have been found with bodies and legs. So trilobites have been hard to classify.

(a) Imagine you are the first to find the remains of a trilobite. It has only a shell and no body or legs. Which invertebrate group might you put it in by mistake?

(b) Why is it hard to classify the trilobites?

(c) Which arthropod group are the trilobites most like? Describe a feature that makes them like one of the other groups.

Questions

1. a What feature do we use to classify invertebrates with no legs?

   b An earthworm has no legs and a soft body with segments. In which group does it belong?

2. a What is an arthropod?

   b What is the difference between a centipede and a roundworm?

3. Cuvier's microscopes were not as powerful as modern ones. Why do you think his classification was different from ours today?

4. *Challenge:* design a key to identify the arthropod groups. Use your key to find out which arthropod group this animal on the right belongs to.

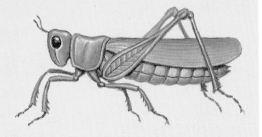

For your notes:

- Invertebrates are classified into seven groups.

- The groups are **jellyfish, starfish, flatworms, roundworms, segmented worms, molluscs** and **arthropods**.

- The arthropod group is divided into **crustaceans, centipedes** and **millipedes, spiders** and **insects**.

D6 The right size

Inuit pen pal

Biork lives in Alaska, close to the Arctic Circle. Biork's ancestors are called Eskimos or Inuit people.

Biork's people have survived the cold Arctic conditions for thousands of years. Biork's grandfather used to go out hunting for seals. He wore clothes made of animal skins and built overnight shelters out of ice to keep warm.

Inuit people are born with short compact bodies. It's a feature that has been passed on through our families.

The Inuit people have short, heavy, compact bodies. Biork has often wondered why she is small. She would like to be tall and thin like her American pen friend in the USA. Biork's grandfather says it's a feature that helps them keep warm.

(a) **Why do you think that Inuit people are small?**

Biork's body shape is inherited from her parents, but it also depends on her surroundings, lifestyle and upbringing.

Lots of variables can affect our height and weight. Some of these are:

● food – Inuits eat mainly seal meat, which is very fatty.

● exercise – if it is very cold, you stay inside. If you move around less, you do not use up so much food.

● seasons – children grow less in the winter because they use more of the chemical energy they get from their food to keep warm.

● illness and stress can make you lose weight.

Research

A group of Canadian scientists studied the heights of more than 150 Inuit children. They compared them with a **sample** of 150 children in the USA. It was important to take as large a sample as possible. If you take a small sample, you might get mainly tall people or mainly small people. If you take a large sample, you are more likely to see the whole **range** of different heights.

b Why do you think it is better to study a sample of 150 children from each place rather than only 10?

c How would you choose the children that you were going to study?

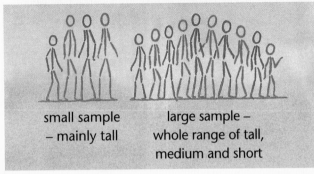

small sample – mainly tall

large sample – whole range of tall, medium and short

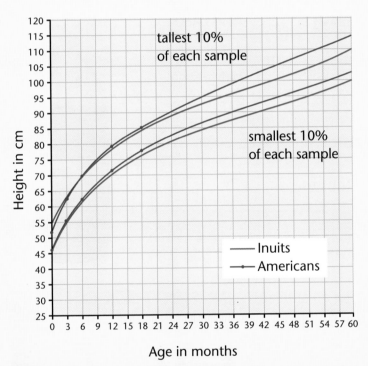

tallest 10% of each sample

smallest 10% of each sample

Height in cm

— Inuits

— Americans

Age in months

Making comparisons

When they had collected their data, the scientists compared the tallest 10% of each sample and the smallest 10% of each sample. They did this to see the differences.

Analysing the results

The graph shows the data for the Inuit girls compared with girls in the USA.

When both groups of children are 12 months old, the lines on the graph for the Inuit children and the American children are close together. This means that their heights are similar. After this, the lines are further apart, showing that the American children are taller.

Questions

Discuss the graph with your partner and use it to answer the questions.

1 Why do you think the scientists took a large sample from each place but only compared the tallest and the shortest in the range?

2 Was there any difference in the heights of the Inuit and the American girls when they were 12 months old?

3 What do you notice about the growth of the Inuit and the American girls between the ages of 12 and 60 months?

4 How do you think the data for boys might differ from the data for girls?

5 The scientists wanted to find out whether the Inuits are small because size is inherited. What other data might they have collected and analysed? Explain why you made your suggestions.

6 Do you think Biork's grandfather was right? Give two reasons for your answer.

7 Could you use your graph to predict the height of the Inuit girls when they are 30?

47

E1 Acids, bases, alkalis

Learn about:
- Acids, bases and alkalis
- Indicators

Acids

An **acid** is a solution of a particular kind of solid or gas in water. An acid that does not contain much water is a **concentrated** acid. An acid that contains quite a lot of water is called a **dilute** acid.

Some acids may be **corrosive**, **toxic**, **harmful** or **irritant**. Hazard warning labels are used on their containers to warn people of the dangers.

Do you remember?

You already know that some solids dissolve in water to make a solution, and others do not.

CORROSIVE
may destroy living
tissues on contact

TOXIC
poisonous

HARMFUL
may have health risk
if breathed in, taken
internally or absorbed
through skin

IRRITANT
non-corrosive
substance which
can cause red or
blistered skin

Useful acids

Some insects and even some plants have acidic stings, which they use to attack animals that try to eat them.

(a) Ants have acidic stings. When do you think ants use their stings?

Sulphuric acid is a useful acid. For example, it is used in car batteries. It is corrosive. You need to handle it carefully to avoid burning holes in your clothes or your skin.

We use products that contain acids every day, and not all of them are dangerous. Many acids are not corrosive – they do not burn or eat things away. This may be because they have been diluted with water. Vinegar is safe. But concentrated ethanoic acid, the acid contained in vinegar, is very corrosive. Acids are less corrosive when you dilute them with lots of water.

Vinegar contains
ethanoic acid.

ENJOY-ICE COLD
LOW CALORIE SOFT DRINK WITH
VEGETABLE EXTRACTS WITH SWEETENER.
INGREDIENTS:
CARBONATED WATER, COLOUR (CARAMEL E 150d),
SWEETENER (ASPARTAME), PHOSPHORIC ACID,
FLAVOURINGS, CITRIC ACID, PRESERVATIVE (E 211).
CONTAINS A SOURCE OF PHENYLALANINE.
Nutrition Information per 100ml
Energy: 1.9kJ, 0.4 kcal
Protein: 0 g
Carbohydrate: 0 g
Fat: 0 g
BEST BEFORE END - see base of can for date
CANNED UNDER AUTHORITY OF THE COCA-COLA
COMPANY BY COCA-COLA & SCHWEPPES
BEVERAGES LTD., UXBRIDGE, UB8 1EZ

Acids in foods

Many of the foods we eat contain acids. They can give food a sour taste.

Some acids have very long names, but the word 'acid' is part of the name.

(b) Look at the pictures here of foods and the acids they contain. Make a list of all the acids you can think of that are found in foods or drinks. What sort of taste do some of them have?

tannic acid

citric acid

lactic acid

ascorbic acid,
which is
vitamin C

Bases and alkalis

Look at the photo showing substances found in the kitchen or bathroom. They all contain **bases**. A base is the opposite of an acid – it cancels out acidity.

Some bases dissolve in water. We call these **alkalis**. Like acids, many alkalis are corrosive, toxic, harmful or irritant.

C Explain what alkalis and bases have in common and how they differ.

Finding out which is which

Using your senses is not a safe way of finding out whether a solution is acidic or alkaline. The best way is to use an indicator. An **indicator** is a coloured substance that shows you whether the solution you are testing is an acid, an alkali or **neutral**. A neutral solution is neither acidic nor alkaline. Pure water is neutral.

You can make an indicator using coloured dyes from flowers, fruits and vegetables. When an indicator is mixed with an acid or alkali, it will change colour.

Using litmus

Litmus is an indicator made from lichens. We use it in laboratories to find out whether solutions are acidic, alkaline or neutral. Acids turn blue litmus paper red. Alkalis turn red litmus paper blue. Both blue and red litmus paper stay the same colour with neutral solutions.

Did you know?

Red cabbage acts as an indicator. It turns bright red in acids, and blue in alkalis.

Questions

1 Explain why battery acid has a corrosive hazard label but cola does not.

2 Lynn and Ryan were arguing. Lynn said that all acids are corrosive. Ryan said they're not. Who was right? Explain your answer.

3 What is an indicator? Explain what indicators are used for.

4 Look carefully at the table which shows some of the indicators you might find in a paint factory laboratory.

Indicator	Acidic	Alkaline	Neutral
phenolphthalein	clear	red	clear
methyl orange	red	yellow	yellow
methyl red	pink	yellow	yellow

Which indicator would show the difference between an alkaline solution and a neutral solution the best? Explain your answer.

For your notes:

● **Acids** may be **corrosive, toxic, harmful** or **irritant**.

● **Bases** are the opposites of acids. They cancel out acidity.

● An **alkali** is a soluble base. Alkalis may be corrosive, toxic, harmful or irritant.

● A **neutral** solution is neither acidic nor alkaline.

● **Indicators** turn different colours with acidic, alkaline and neutral solutions.

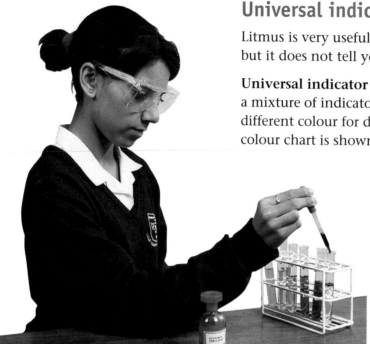

Universal indicator

Litmus is very useful for telling whether a substance is acidic or alkaline, but it does not tell you how strong or weak it is or how dangerous it is.

Universal indicator is another indicator made from plants. Because it is a mixture of indicators it can produce a range of colours. It will show a different colour for different strengths of acidic or alkaline solutions. The colour chart is shown on the next page.

- Universal indicator comes in liquid or paper form.

- To use universal indicator liquid, add two drops to the solution you want to test.

- To use universal indicator paper, dip a glass rod into the solution. Touch the indicator paper with the glass rod.

- Compare the colour with the chart.

a Give two reasons why universal indicator is more useful than litmus.

The pH scale

Some acidic solutions are strongly acidic, and others are weakly acidic. The same is true for alkaline solutions. We use pH numbers to measure the strength of the acidity or alkalinity. (*Remember:* write it with a small 'p' and capital 'H'.)

A neutral solution has pH 7. Acidic solutions have a pH less than 7. The lower the pH, the stronger the acidity.

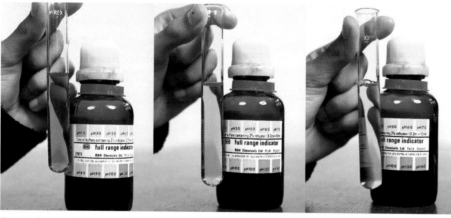

Strongly acidic.　　Less strongly acidic.　　Weakly acidic.

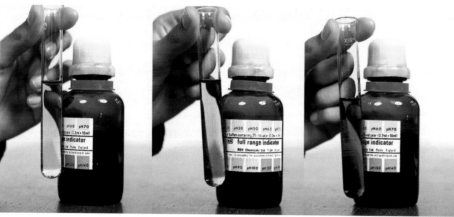

Neutral.　　Weakly alkaline.　　Strongly alkaline.

Look at the **pH scale** below. Hydrochloric acid is strongly acidic, with pH 1. Vinegar is more weakly acidic, with pH 3–4. Alkaline solutions have a pH higher than 7. Sodium hydroxide solution is strongly alkaline, with pH 14. Ammonia is more weakly alkaline, with pH 11.

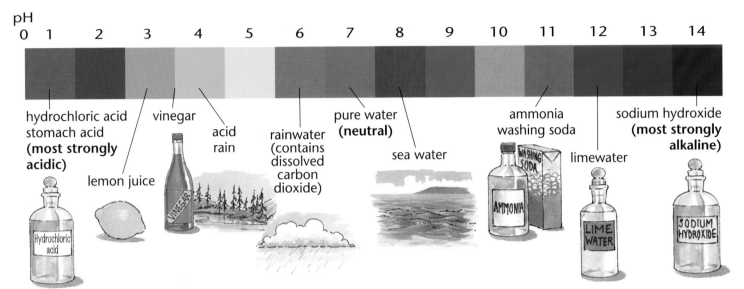

b The pH of milk is 6.5. What colour would universal indicator turn with milk?

c What effect do you think diluting an acid will have on its pH?

Questions

1 Look very carefully at the pH chart above. Copy the table below and complete it. The first line has been done for you.

Liquid	Colour with universal indicator	pH	Description
salt water		8	weakly alkaline
rainwater			
soap			
lemon juice			

2 During a long-distance space mission, one of the astronauts is suspected to have an unusual viral illness. The virus produces an acidic substance in the saliva.

Explain how you could use universal indicator paper to check whether any of the other crew members have contracted the disease.

3 Why would universal indicator be more useful than litmus if you were working in a food technology laboratory?

4 Design an illustrated flow chart for a technician which shows at a glance how to use universal indicator paper.

Did you know?

Alkaline solutions have a high pH and they prefer to react with acids. This is because all acids contain hydrogen. *Remember*:

● **pH** '**p**refers **H**ydrogen'.

For your notes:

● **Universal indicator** turns different colours with different strengths of acidity and alkalinity.

● An acidic solution may be strongly or weakly acidic. An alkaline solution may be strongly or weakly alkaline.

● The **pH scale** is used to measure the strengths of acidic and alkaline solutions.

● Water is neutral, so it has pH 7.

E3 Taking away acidity

Learn about:
● Neutralisation

Putting acids and bases together

We say that a base is the opposite of an acid, because if you add a base to an acid you take away its acidity and make a new substance.

If you add vinegar to the base sodium hydrogencarbonate, you will see bubbles. The vinegar, which is an acid, is used up. The pH falls and you get a neutral solution. This is called **neutralisation**.

Remember, a base that will dissolve in water is called an alkali. So all alkalis dissolve in water, and they also neutralise acids.

Acid indigestion

Your stomach makes hydrochloric acid, which helps you digest your food. The stomach has a special layer which stops the acid corroding your insides! Sometimes the stomach produces too much acid and you might get acid indigestion. This happens because you have eaten too much, not because you have eaten acidic food.

Indigestion remedies are sometimes called 'antacids' or 'anti-acids'. They have alkalis or bases in them to neutralise the stomach acid.

Do you remember?

When you mix materials, this often causes them to change.

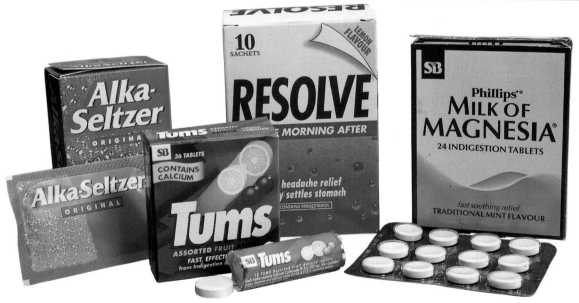

Toothpaste

Inside your mouth, there are lots of bacteria which feed on the food remains left on your teeth. These bacteria produce acid. Your teeth are covered with a substance called enamel, which is the hardest substance in your body, but it can still be attacked by acids. Toothpaste is a base. It helps to neutralise the acids produced by bacteria and acids in your food.

Did you know?

Before all these medicines were available, people just used to take a spoonful of sodium hydrogencarbonate in a glass of water. It was cheap and it worked!

Problem soil

Some soils are slightly acidic. This is fine for plants such as heathers, which grow in places like the Yorkshire moors. The rhododendrons in the photo also like acidic soil. But most plants and food crops will not grow in acidic soil. The acid stops the bacteria in the soil breaking down the dead plants properly. This prevents the water from draining away and the soil becomes waterlogged. Peat bogs have very wet, acidic soil. Farmers and gardeners dig in a base called **lime** to neutralise the acidic soil.

Sting relief

A bee sting is acidic. You can relieve the pain with sodium hydrogencarbonate or calamine lotion, which will neutralise the sting.

A wasp sting is alkaline.

a Which substances would you find at home to treat a wasp sting?

Industrial effluent

Water from factories is sometimes contaminated with acids or alkalis. At the water treatment works, a stream of alkaline water may be mixed with a stream of acidic waste water so that they neutralise each other. Lime may be added to acidic waste water to neutralise the acid, and then left to settle out. Alternatively, the acidic water is passed over limestone, as this diagram shows.

Alkaline water is neutralised by adding acid, or bubbling carbon dioxide through it to form carbonic acid.

b What could happen if acids or alkalis from factories leaked into our rivers?

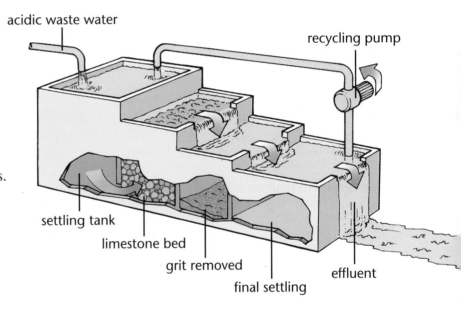

acidic waste water

recycling pump

settling tank

limestone bed

grit removed

final settling

effluent

Questions

1 Explain why alkalis are often described as the opposites of acids.

2 What happens when sodium hydrogencarbonate is added to vinegar? What do you see?

3 Val had acid indigestion. Roger told her not to drink cola. Why do you think he did this? What should she do?

4 A manufacturer wants to develop a range of fruit-flavoured toothpastes. Do you think this is a good idea? Explain your answer. (*Remember:* fruits are acidic.)

5 What would you do to neutralise an acidic soil? How would you test to see if the neutralisation has worked?

6 How is alkaline industrial waste water neutralised?

For your notes:

● When you add a base to an acid, a change called **neutralisation** takes place.

53

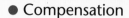
E4 Problem soil

Soil acidity

James Beck is worried about the acidic soil on his farm. He has decided to use 'Superbase' to neutralise the soil. He wants to use enough base to neutralise all the acid in the soil, but not too much because that would leave the soil alkaline. It would also waste money.

Ask an expert

James has asked soil scientist Sarah Jones to advise him.

*I will mix a **sample** of soil with water and then measure the number of acid particles dissolved in the water. Then I can estimate the amount of base we need*

Sarah finds that to neutralise the acid particles, James must add the same number of 'Superbase' particles. This diagram shows this.

acid particles in the soil

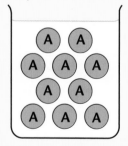

same number of base particles needed to neutralise acid particles

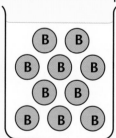

Sarah calculates that one bag of 'Superbase' dissolved in 10 litres of water will neutralise the soil in the field.

ⓐ **How reliable do you think Sarah's calculation is?**

James is still keen to save money. He decides to add another 10 litres of water to Sarah's solution. He then has 20 litres of 'Superbase' solution. He uses 10 litres of it now and saves the other 10 litres for next time.

ⓑ **Do you think this will make any difference to the amount of soil he can neutralise?**

False economy

Sarah was surprised when she found that James' soil was still acidic.

'Where did I go wrong?' wondered Sarah. James told her what he did.

*If you add twice as much water, you must **compensate** by adding more base. Or you can use twice as much of your solution.*

My solution had the right number of base particles to neutralise the acid particles in the soil. When you added another 10 litres of water, you still had the same number of base particles because you only used one bag of 'Superbase'. So the 10 litres of solution you used had only half as many base particles as my 10 litres.

Sarah's solution   water   James' solution   James uses 10 litres

The table shows some examples of solutions that James could use.

Volume of water added to one bag of base	Dilution factor compared with Sarah's solution	Use
20 litres	2	2 times the amount
40 litres	4	4 times the amount
100 litres	10	10 times the amount
200 litres	20	20 times the amount
5 litres	0.5	0.5 times the amount

c If the volume of water is 120 litres, what is the dilution factor?

d Calculate the dilution factor if James added 390 litres of water to Sarah's 10 litres of base solution.

Questions

Discuss these questions with your partner. Think about **compensation**. Write down your answers.

1 If you add more water to Sarah's base solution, what will happen?

2 What will you have to do to neutralise the same amount of soil with your new solution?

3 How would you explain 'compensation' to James?

4 If the dilution factor for the base solution is 5, how much more of the solution will be needed to compensate for the dilution?

5 James has been convinced that he cannot save money by diluting the base solution, so he has decided to look for a cheaper alternative. Should he simply look at price? What else should he think about? Look back at the spreads in this unit to give you some ideas about the different properties of bases.

F1 Changing materials

Physical and chemical changes

When you boil water there is a **physical change**. But no new substance has been made. It is water at the beginning and water at the end.

Most irreversible changes, like burning a match, involve making new substances. The substances in the wood have gone, and in their place are new substances in the air and in the ash. Changes that involve making new substances are called **chemical changes**.

ⓐ Describe another: (i) physical change (ii) chemical change.

Chemical reactions

Every time a new substance is made, there is a **chemical reaction**. A chemical change, like burning a match, may involve lots of chemical reactions.

Spotting a chemical reaction

The following are signs that a chemical reaction is happening.

1 *A new substance being made.* We can see that the car has rusted because it has turned orange. A new substance, rust, has formed. When the tablet is put into water, bubbles are made. This is because a new substance, a gas, is being made. The glue sets because a new, solid substance is made.

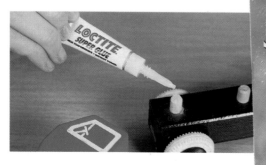

2 *A substance being used up.* In the end, the iron in the car will all have gone. Similarly, the white powders in the tablet will have gone, as will the clear, colourless liquid in the tube of glue.

3 *Energy changes.* Another sign of a chemical reaction is energy being given out or taken in. The fireworks give out energy as sound, light and heat when they explode. The explosion is a chemical reaction.

ⓑ Give three reasons why we think burning a match involves chemical reactions.

Reactants and products

Chemical reactions make new substances. These new substances are called **products**. The substances you started with are called **reactants**. Reactants are used up during the reaction.

We are going to look at a very simple chemical reaction. This photo shows iron being heated with sulphur. The information in the table shows the observations that were made during the reaction.

We started with …		A chemical reaction!	We ended up with …
a grey powder that was magnetic. This was iron.	a yellow powder that was not magnetic. This was sulphur.	The substances in the test tube glowed bright red.	a grey solid that was not magnetic. This was iron sulphide.

C Use the information in the table to answer these questions.
(i) What were the two reactants? (ii) Name the product.
(iii) Give three reasons why you think a chemical reaction took place.

Word equations

We use **word equations** to describe chemical reactions.

iron + sulphur	$\rightarrow$	iron sulphide

The substances we start with – the **reactants** – are put on the left.

The chemical reaction is shown by an arrow.

The new substances – the **products** – are put on the right.

Word equations are useful because they sum up what you know about the chemical reaction in very few words.

Questions

1 Look at this photo. It shows a sparkler burning. A scientist describes what is happening:

The sparkler is covered in a substance called 'iron'. There is another substance in the air called 'oxygen'. Iron and oxygen are used up during the reaction. The new substance made is 'iron oxide'.

a What evidence is there in the photo for a chemical reaction?

b Read what the scientist says. What other evidence is there that a chemical reaction is happening?

c What are the two reactants in this chemical reaction?

d What is the product of this reaction?

e Write a word equation for this reaction.

2 This photo shows two colourless, clear liquids being mixed. Do you think a chemical reaction is happening? Give two reasons for your answer.

For your notes:

- A **chemical reaction** is a change that makes a new substance.

- Colour changes, bubbles and a solid forming are signs that a new substance is being made in a chemical reaction.

- Energy being taken in or given out also suggests a chemical reaction may be happening.

- In a chemical reaction, the substances that are used up are **reactants** and the new substances are **products**.

- **Word equations** are used to show chemical reactions.

F2 **Acids and metals**

Learn about:
- Acids reacting with metals
- A test for hydrogen
- Corrosion

Hubble, bubble!

If you add a few granules of zinc to some acid in a beaker, it fizzes. The photo on the right shows zinc being added to sulphuric acid. The zinc is used up and bubbles form. The temperature of the mixture rises. When all of the zinc has reacted, the bubbles stop.

a **How can you tell that there is a chemical reaction between zinc and sulphuric acid?**

A gas is made during this chemical reaction. The gas makes the bubbles. You can collect the gas in a test tube. You have to start with the upside-down tube full of water. The gas pushes the water out of the test tube.

The gas is called **hydrogen**. Hydrogen is one of the products of this chemical reaction. The other product is much harder to detect. It is a soluble, colourless substance called zinc sulphate.

b **For this reaction, name:**
(i) the reactants (ii) the products.

c **Copy and complete this word equation for the reaction.**

_____ _____ + _____ → _____ + _____ _____

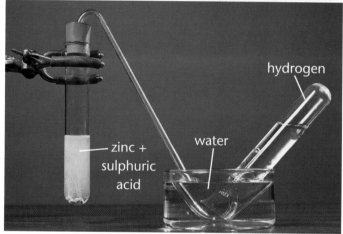

hydrogen

zinc + sulphuric acid

water

PoP

A test for hydrogen

Once you have a test tube full of gas, you can test it by putting a lighted splint near the top. This gas is explosive, so you will hear a 'pop'! Hydrogen is the only gas that pops like this.

This test for hydrogen is another chemical reaction. You can tell because energy is given out – the pop sound.

Finding the zinc sulphate

The zinc reacts with the sulphuric acid. The hydrogen gas that is made mixes with the air. You are left with a clear, colourless solution.

Look at this photo. If you heat this solution, the water will evaporate. You end up with a white solid. That is the zinc sulphate.

Naming substances

Where does the name 'zinc sulphate' come from? You have probably worked out for yourself that the 'zinc' part comes from one of the reactants, zinc, and the 'sulphate' part is something to do with the other reactant, sulphuric acid.

Sulphuric acid always forms **sulphates** when it is neutralised. You met **neutralisation** in Unit E Acids and alkalis. Neutralisation happens when the 'acidity' is taken away from the acid. Now you know why. The acid has been used up in the chemical reaction.

Other acids and metals

Other acids also react with metals to make hydrogen. For example, hydrochloric acid reacts with zinc to make hydrogen and zinc chloride. During neutralisation, hydrochloric acid makes chlorides.

Many metals, other than zinc, react with acids, including aluminium and magnesium.

d Copy and complete this word equation.

hydrochloric acid + zinc → _____ + _____ _____

e Write your own word equation for the reaction between hydochloric acid and aluminium.

Corrosion

The sulphuric acid seemed to eat away the zinc during the chemical reaction. The zinc was used up as the hydrogen was made. When metals are used up by chemical reactions we call it **corrosion**.

The photo on the right shows a metal tray that a car battery sits on. The tray has been corroded by acid escaping from the car battery.

Did you know?

Acids have names ending in -ic. You have already met sulphuric acid and hydrochloric acid. The acid in vinegar is ethanoic acid and the acid in lemon juice is citric acid. Other acids include nitric acid, phosphoric acid and ascorbic acid (vitamin C).

Questions

1 Describe how you would test a gas to see if it was hydrogen.

2 An iron nail is put into sulphuric acid. Bubbles form.

  a What are the two products of this reaction?

  b Write a word equation for this reaction.

3 Look through the information on these two pages.

  a Identify one physical change that is described. Explain how you knew it was a physical change.

  b Identify four chemical changes that are described. Explain how you know each is a chemical change that involves one or more chemical reactions.

4 **a** What is corrosion?

  b Why do we want to stop chemical reactions that cause corrosion?

For your notes:

- Some metals react with acids to make **hydrogen** gas. This is one type of **neutralisation**.

- Sulphuric acid forms **sulphates** when it is neutralised.

- Hydrogen gas pops when you test it with a lighted splint.

- Metals **corrode** because of chemical reactions.

F3 **Acids and carbonates**

Learn about:
● Acids reacting with carbonates
● A test for carbon dioxide
● Carbonates

Fizz!

If you take a piece of chalk and drop it into acid, as shown in this photo, you will see bubbles of gas. You also see the insoluble chalk being used up during the reaction.

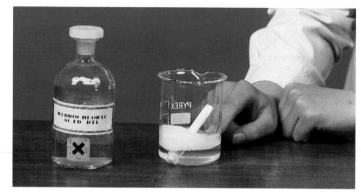

The gas is **carbon dioxide**. Like hydrogen, carbon dioxide is a colourless gas. Unlike hydrogen, carbon dioxide does not burn. If you put a lighted splint into a test tube of carbon dioxide, it will go out.

A test for carbon dioxide

You test for carbon dioxide by bubbling the gas through **limewater**. Look at the photo on the right. Carbon dioxide makes limewater go cloudy.

a How do you know that there is a chemical reaction when:
   (i) chalk is put into acid?
   (ii) carbon dioxide is bubbled through limewater?

Another neutralisation

Chalk is calcium carbonate. Look at the photo of the chalk reacting with the acid. The acid is hydrochloric acid. This is a neutralisation reaction. The hydrochloric acid is used up during the reaction.

One of the products of this reaction is carbon dioxide. Another product is a soluble, colourless substance called calcium chloride. The third product is water.

b Copy and complete this word equation for the chemical reaction.

____ ____ + ____ ____ → ____ ____ + ____ ____ + ____

> ### Did you know?
>
> Hydrochloric acid always makes **chlorides** when it is neutralised, in the same way as sulphuric acid always makes sulphates when it is neutralised.

Carbonates

There are many **carbonates**, including calcium carbonate, copper carbonate, sodium carbonate and sodium hydrogencarbonate. All carbonates make carbon dioxide when they neutralise acids.

Light and fluffy

We use the reaction between acids and carbonates when we make a sponge cake. A good sponge cake is light and fluffy, with lots of tiny holes. These holes are made by bubbles of carbon dioxide gas.

The carbon dioxide gas is made by mixing baking powder into the cake mixture. Baking powder contains an acid and a carbonate. They react together when the baking powder is added to the cake mixture. Bubbles of carbon dioxide are made.

The reaction does not start in the packet because both the acid and the carbonate are solids. They only react when they are dissolved. Dissolving them starts the reaction.

C The carbonate in baking powder is 'bicarb'. Bicarb is used to make soda bread rise. Explain why the soda bread is fluffier when you include: (i) lemon juice or (ii) sour milk in the bread mixture.

Carbonate rocks

Limestone, marble and chalk are common rocks that are made from calcium carbonate. Like all carbonates, these rocks will react with acids in the surroundings. Air pollution can make the rainwater very acidic. We call this acidic rainwater **acid rain**. Acid rain reacts with rocks made from calcium carbonate and the rock is slowly used up.

We make buildings and statues out of limestone and marble. They react with acids in the surroundings, particularly if the rainwater is acidic. The photo on the right shows the Sphinx, a statue from ancient times. The surface has been used up in chemical reactions.

Malachite is another carbonate rock. It is made from copper carbonate and has a distinctive green colour. When polished, malachite makes beautiful jewellery.

Questions

1 You are given three test tubes containing samples of hydrogen, carbon dioxide and air. Plan an experiment to find out which is which.

2 Use the information on these two pages and on the previous two pages to work out the products of the following chemical reactions. Copy and complete the word equations to show the products.

   a sodium carbonate + hydrochloric acid →

   b calcium carbonate + sulphuric acid →

   c zinc + hydrochloric acid →

3 Calcium carbonate reacts with hydrochloric acid to make calcium chloride, water and carbon dioxide. Explain how the soluble, colourless calcium chloride could be separated from the solution at the end of the reaction. (*Hint:* look back at page 58 if you get stuck.)

4 Ellen was buying a malachite ring. During her conversation with the jeweller she mentioned that she was studying chemistry. The jeweller suggested that she never wear the ring in the laboratory. Explain why.

For your notes:

- Acids react with **carbonates** to make **carbon dioxide**. This is a neutralisation reaction.

- Carbon dioxide gas turns **limewater** milky.

- Limestone, marble and chalk are made from calcium carbonate, so they react with acids. This means that they wear away because of acids in the surroundings.

F4 Firefighting

Oxygen supply

Is air needed for burning? Look at the picture. Three jars of different sizes are placed over a burning candle.

(a) How does the amount of air available to the candle affect how long it burns?

The candle under the biggest jar burned for the longest time. This jar has the most air in it. The candle needs a substance in air for burning. When that substance in air is used up, the candle goes out.

The substance in the air that is used up during burning is **oxygen**.

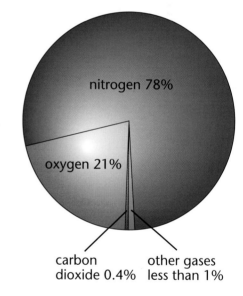

Candle went out after 10 seconds.

Candle went out after 8 seconds.

Candle went out after 4 seconds.

(b) Look at the pie chart. How much of the air is oxygen?

(c) Imagine the jars were filled with pure oxygen instead of air. Predict how long the candle under the largest jar would have burned for.

You may have predicted 50 seconds, because the air is about one-fifth oxygen. Pure oxygen would have five times the amount of oxygen as air. In fact, the candle would not stay alight this long. It would burn brighter and faster in the pure oxygen.

Combustion

Burning is a chemical reaction. The reactants are the material that burns, called the **fuel**, and oxygen. The fuel and the oxygen are used up during the chemical reaction.

The scientific name for burning is **combustion**.

nitrogen 78%

oxygen 21%

carbon dioxide 0.4%

other gases less than 1%

Firefighting

Dangerous fires start when burning gets out of control. They produce large amounts of energy as heat and light. The fire in the building uses up the oxygen, which adds to the danger.

Our bodies need a constant supply of oxygen to stay alive. People trapped in a burning building cannot breathe. Look at the photo. Firefighters sometimes have a cylinder of oxygen on their backs, so they can breathe in a burning building.

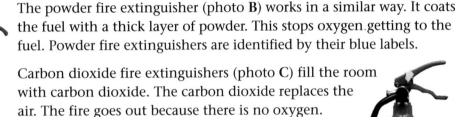

Fires need fuel and oxygen. They also need a little bit of energy to get them going. Matches do not burst into flames on their own – you have to strike them. You need a spark to light a gas cooker or a Bunsen burner. Once the fire has started, the energy given out is enough to get the next bit of fuel going. The fire spreads.

We represent this as a **fire triangle**. A fire triangle is useful because it reminds us that there are three ways of putting out a fire:

● remove the fuel

● remove the oxygen

● cool it down so there is not enough energy to start the next bit of fuel burning.

A

Photos **A** to **C** show equipment use to put out small fires.

The fire blanket (photo **A**) is draped over the fire. It stops fresh air getting to the fuel. The oxygen in the air under the blanket is used up and the fire goes out.

The powder fire extinguisher (photo **B**) works in a similar way. It coats the fuel with a thick layer of powder. This stops oxygen getting to the fuel. Powder fire extinguishers are identified by their blue labels.

Carbon dioxide fire extinguishers (photo **C**) fill the room with carbon dioxide. The carbon dioxide replaces the air. The fire goes out because there is no oxygen. Carbon dioxide extinguishers are identified by their black labels and the letters 'CO_2'.

B

Water fire extinguishers work because they cool down the fuel. The energy goes into heating up the water and there isn't any left to get the next piece of fuel burning. Water fire extinguishers have red labels.

C

Questions

1 Describe the evidence that burning is a chemical reaction.

2 Parts **a** to **d** are four ways of putting out a fire. Explain how each works, using the fire triangle:

   a digging a 'firebreak' (removing all the trees and buildings)

   b using a carbon dioxide fire extinguisher

   c covering with a fire blanket

   d soaking the fuel in water.

3 Look at the photo of the candle burning. The candle wax melts and then soaks into the wick. The liquid candle wax then burns.

   a What physical change is taking place? Explain why you think this is a physical change.

   b What chemical reaction is taking place? Explain why you think this is a chemical change.

For your notes:

● Burning is a chemical reaction. Another word for burning is **combustion**.

● Combustion uses up **oxygen** and **fuel**.

● For a fire to start it needs fuel, oxygen and a little energy (a spark). You can put out a fire by removing one of these three things.

Products of combustion

What are the products of combustion? It depends on which fuel you use.

One of the simplest fuels is **charcoal**. We use charcoal on our barbecues, like the one in the photo. Charcoal is almost pure carbon. When you burn carbon you make carbon dioxide.

carbon + oxygen → carbon dioxide

Burning metals

As well as fuels, we can also burn metals. The metal magnesium burns with a bright white flame. The magnesium reacts with oxygen in the air. A white powder is made. This is magnesium oxide.

magnesium + oxygen → magnesium oxide

Oxides

Carbon dioxide and magnesium oxide are both products of combustion. In fact, all the products of combustion are **oxides**.

Oxides are made when oxygen reacts with other substances in a burning reaction. The name oxide comes from the word oxygen. We take the first two letters of oxygen and add 'ide'.

a **What do all oxides have in common?**

Other fuels

Another fuel is hydrogen. When hydrogen burns, it makes an oxide. You would expect the oxide to be called 'hydrogen oxide' but 'hydrogen oxide' turns out to be water.

hydrogen + oxygen → water

Many fuels contain both carbon and hydrogen. We call these fuels **hydrocarbons**. When hydrocarbons burn, both carbon dioxide and water are made.

Methane is the fuel in natural gas. Methane is a hydrocarbon.

methane + oxygen → carbon dioxide + water

Candle wax, petrol and camping gas are all examples of hydrocarbons.

b **Propane is another hydrocarbon. Write a word equation for the combustion of propane.**

candle

camping stove

petrol

Making predictions

Carl and Emma were burning substances. They were trying to predict the products of combustion.

First they burned copper. This photo shows copper being burned. A black substance forms on the surface.

It will make copper oxide.

c Write a word equation for copper burning.

d Predict the product when aluminium burns. Write a word equation describing your prediction.

Then they watched their teacher burning sulphur. Look at the photo on the right. Sulphur is a yellow substance. It burns with a blue flame.

It will make sulphur oxide.

Oxide or dioxide?

Carbon can burn to make carbon monoxide or carbon dioxide. These are both colourless gases with no smell.

Whether you get carbon dioxide or carbon monoxide depends on the oxygen supply. If there is lots of oxygen, you make carbon dioxide. If there is limited oxygen, you get carbon monoxide.

Carbon monoxide is **toxic**. That means that it poisons humans. It is also **lethal**. That means it can kill you.

We call it sulphur dioxide.

Questions

1 Carbon monoxide is a dangerous substance.

  a Why is carbon monoxide dangerous?

  b When is carbon monoxide formed rather than carbon dioxide?

  c Why should you always make sure there is a good supply of fresh air when burning a hydrocarbon fuel?

2 Use the information in 'For your notes' to copy and complete these sentences.

  a Oxides are made when …

  b Hydrocarbons burn to make …

  c Carbonates neutralise acids and make …

  d Metals neutralise acids and make …

  e When sulphuric acid is neutralised you make …

  f When hydrochloric acid is neutralised you make …

3 Write word equations for:

  a zinc burning   **b** magnesium neutralising hydrochloric acid

  c ethane (a hydrocarbon) burning.

For your notes:

- **Oxides** are made during burning. They are the products of combustion.

- Knowing about groups of substances such as oxides helps you make predictions about chemical reactions.

- Other groups of substances include sulphates, carbonates, acids, chlorides and **hydrocarbons**.

- Hydrocarbons burn to make both carbon dioxide and water.

F6 Getting hotter

Fuels in industry

We burn large amounts of fuels such as coal, oil and gas to give us light energy, heat energy and to make things move. The more fuel we burn, the more carbon dioxide we release into the air.

The greenhouse effect

The carbon dioxide in the air has the same effect as the glass in a greenhouse. The glass stops some of the heat energy in the greenhouse escaping, and the plants stay warm. Carbon dioxide stops some of the heat energy from the Earth escaping, and the Earth stays warm. This is called the **greenhouse effect**.

Scientists think that too much carbon dioxide will increase the greenhouse effect and the Earth will get too hot. They think that there is a relationship between the level of carbon dioxide in the air and the temperature of the Earth. As the level of carbon dioxide rises, the Earth gets hotter.

The biodome experiment

A team of students decided to find out whether there is a relationship between the temperature of the Earth and carbon dioxide levels in the atmosphere. It is difficult to measure large-scale changes around the Earth, so they designed a **model** to represent the Earth.

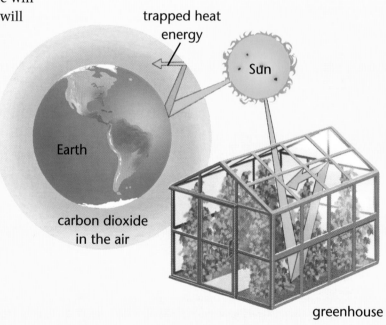

trapped heat energy

Sun

Earth

carbon dioxide in the air

greenhouse

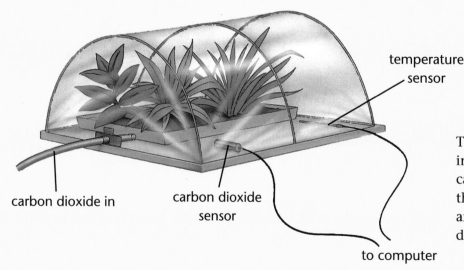

carbon dioxide in

carbon dioxide sensor

temperature sensor

to computer

They set up a biodome, as shown in this picture. Carbon dioxide can be added through the tube at the side. The carbon dioxide level and the temperature inside the dome are recorded by a computer.

Any relationship?

a Can you identify the *input* and *outcome variables* in the biodome experiment?

We use the word **relationship** to describe how the outcome variable changes when we change the input variable.

The students ran the experiment for five days. They increased the level of carbon dioxide each day to see if there was a relationship between carbon dioxide level and temperature. The graph on the computer screen looked like this.

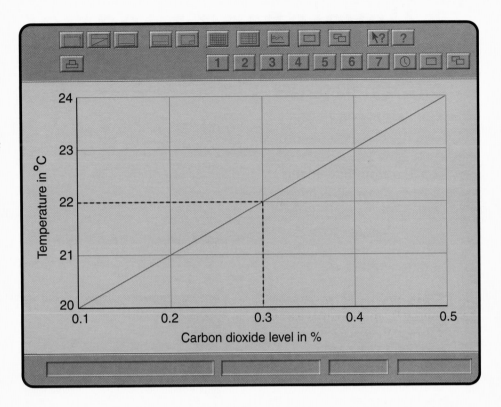

Describing the graph

A relationship describes how the outcome variable changes when the input variable is changed. It describes the pattern shown by a graph.

The graph of temperature against carbon dioxide level shows a relationship. On the first day, the carbon dioxide level went up by 0.1% and the temperature went up by 1°C. This happened every day. The carbon dioxide level and the temperature both went up together. This gives a straight-line graph.

b Look at the graph. What was the temperature in the biodome when the carbon dioxide level was 0.3%? Follow the line up from 0.3% on the *x*-axis and stop when it touches the graph. Then follow the line across to the *y*-axis to find out the temperature.

If the students had carried out the experiment for a few more days, the line of the graph would have continued to show the same pattern.

Questions

Discuss the questions with a partner. Jot down your answers.

1 What do you think the temperature in the biodome would have been when the carbon dioxide level was 1.0%? (You are making a prediction.)

2 Do you think there is a relationship between carbon dioxide levels and the temperature of the Earth? Why do you think that?

3 Use the results from the biodome to predict what you think will happen to the temperature of the Earth over the next 10 years.

4 What do you think the scientists should do next?

5 Find out how the greenhouse effect could affect:

   a polar ice-caps   b weather

   c world food supplies.

G1 Developing theories

Different materials

Matter is the stuff around you. It makes up **materials**, such as water, air and this paper in front of you. But why do materials behave so differently? Why can they flow or float or stay rigid? Why can water do all three at different times?

Ideas about matter

About two and a half thousand years ago in Ancient Greece, groups of philosophers, or thinkers, decided that the world around them could be studied, understood and explained. One of the things they wanted to explain was matter.

Between about 600 and 300 BC, two competing ideas about matter were developed.

The four elements   Idea 1

- Around 580 to 570 BC, the philosopher Thales suggested that everything was made up of water. This may seem weird to us, but it is easy to see how the idea came about. Water can be a **solid**, a **liquid** or a **gas**, so Thales concluded that all solids, liquids and gases are made of water.

Aristotle.

- About 50 years later, Anaximander suggested that everything was made from a single type of material that varied in its hotness, coldness, dryness and wetness.

- Another 70 years later Empedocles had the idea that everything was made up four elemental substances – fire, air, water and earth – mixed in different amounts. Substances changed because there were two forces that could affect them. Empedocles thought these two forces were love and strife.

- The idea of four elemental substances really caught on. A very famous philosopher called Aristotle believed this idea. For the next two thousand years anyone who was educated was taught that Aristotle was right, and everything was made from fire, air, water and/or earth.

Atoms   Idea 2

- In about 480 to 470 BC the philosopher Anaxagoras came up with a very different idea. He suggested that everything was made from tiny 'seeds'.

- About 20 years later Leucippus introduced the idea of **atoms**. His atom was a piece of matter that could not be destroyed or divided.

- This idea was developed in about 430 BC by his student, Democritus, who explained how everything could be made of atoms. He suggested that there were different types of atom that made up different materials. He said that solids were made of atoms that were stuck together, while in liquids and gases the atoms were not stuck together. He suggested that heavier materials, like lead, had bigger atoms.

- However, the idea of atoms was unpopular for the next 2000 years. Only a few people believed it and wrote about it. The most important was Epicurus, who championed the idea about a century after Democritus.

- In AD1649 a scholar called Pierre Gassendi studied what Epicurus had written and republished the idea.

Democritus.

Matter today

In 1803 AD John Dalton published his 'Atomic theory of matter'. A **theory** is a set of ideas that explains something. Dalton had been studying chemical reactions. He realised that his observations could only be explained if there were many 'elemental substances' or **elements**, each made up of a different type of atom. The picture shows Dalton's atoms.

Today atoms are no longer a theory. We have seen them using very powerful microscopes. The scientist who took this photo below even a managed to arrange the atoms to make a stick figure!

John Dalton.

ELEMENTS

⊙	Hydrogen	1	✛	Strontian	46
⊖	Azote	5	✴	Barytes	68
●	Carbon	54	Ⓘ	Iron	50
○	Oxygen	7	Ⓩ	Zinc	56
☮	Phosphorus	9	Ⓒ	Copper	56
⊕	Sulphur	13	Ⓛ	Lead	90
⊙	Magnesia	20	Ⓢ	Silver	190
⊗	Lime	24	Ⓓ	Gold	190
⊖	Soda	28	Ⓟ	Platina	190
⊜	Potash	42	✺	Mercury	167

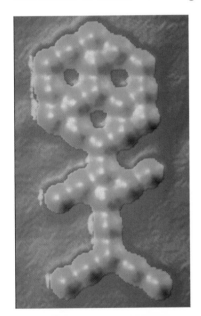

Atoms and particles

Although all substances are made of atoms, those atoms can be arranged in different ways. They can be single atoms, or groups of atoms called **molecules**. We are going to use the word **particle** because that covers both atoms and molecules.

Questions

1 In ancient times there were two different ideas about matter. One was that matter was made up of fire, air, earth and water. The other was that matter was made up of atoms.

   a Use the information on these two pages to write a time line showing how each of these ideas developed.

   b Which idea dominated for two thousand years?

   c Which idea fits with modern scientific theory?

2 Leucippus and Empedocles lived at the same time, although in separate parts of Europe. Write a story about the two of them meeting. Include what you think they would say to each other.

For your notes:

- Ideas about **matter** have changed over time.

- All matter is made of **particles**.

- 'Particle' can mean an **atom** or a group of atoms.

Learn about:
- Particles
- Density

A closer look

The theory that particles make up all materials is called the **particle model**. The particle model explains why solids, liquids and gases behave in different ways. Why does a solid keep its shape? Why is a liquid runny?

Solids

In a solid, the particles are packed closely together in a regular pattern. The particles stay in their places. They are held together by strong forces between the particles.

The particles in a solid vibrate on the spot. Think about aluminium. It is usually a solid because it melts at 660°C. At 0°C the particles vibrate but at 600°C the particles have more energy and vibrate more.

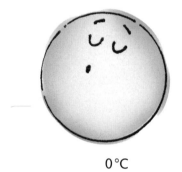

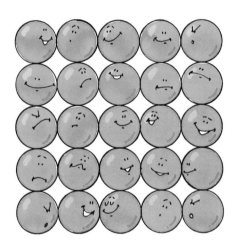

0°C   600°C

Liquids

In a liquid, the particles are still close together. But they are held together by weaker forces and so the particles can move about more. The particles slide over each other. As the particles are always moving, they are not in a regular pattern.

The particles in a liquid still vibrate. Think about aluminium again. It is a liquid between 660°C and 2567°C. At 2000°C the particles will vibrate more than at 700°C.

700°C   2000°C

 a Are the forces between particles stronger in a liquid or in a solid?

Gases

In a gas, there are no forces holding the particles together and they are moving quickly. This means that the particles are very far apart and not arranged in any regular pattern.

Even though they are whizzing about, the particles in a gas still vibrate. Particles of aluminium at 3000 °C will vibrate less than aluminium particles at 4000 °C.

Gases can be poured.

3000 °C

4000 °C

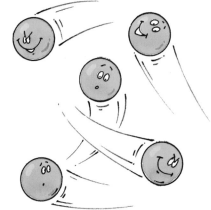

Density

By weighing equal volumes of substances you can compare their **densities**. Liquids and the solids are much **denser** than the gases. Each 1 cm^3 of solid or liquid weighs much more than 1 cm^3 of gas. This is because the particles in a solid and in a liquid are touching. The particles in a gas are far apart. There is a lot of nothing in a gas!

b Are gases more dense or less dense than solids or liquids?

Questions

1 Copy and complete this table choosing from the words below.

large   none   low   medium   high   small

Properties	Solid	Liquid	Gas
force between particles			
regular pattern of particles			
space between particles			
compressibility ('squashability')			
density			
speed of particles			
vibration of particles			

2 One litre of water produces 1333 litres of steam. Explain why there is such a big increase in volume.

For your notes:

- In a solid: the particles vibrate on the spot, they are held together by strong forces, and are closely packed in a regular pattern.

- In a liquid: the particles move about and slide over each other, they are held together by weaker forces, and are closely packed but not in a regular pattern.

- In a gas: the particles are moving quickly. There are no forces holding the particles together. They are far apart in no regular pattern.

- Particles vibrate more at higher temperatures.

- Solids and liquids have greater **density** than gases.

G3 Looking at evidence

Learn about:
- Evidence for the particle model
- Heating and expansion

Solids

How does the particle model fit with what we know about solids?

Solids are strong.

Solids keep their shape.

Solids keep their volume, as long as you don't heat or cool them.

That's because the particles are joined together by strong pulling forces.

That's because the particles are touching and in a regular pattern. Every particle stays in its place because it's joined to the next one.

That's because the pulling forces keep the particles closely packed. They can't get further apart or closer together.

a Solids do not squash. Explain this using the particle model.

Liquids

How does the particle model fit with what we know about liquids?

I can stir a liquid but not a solid.

That's because the particles in a liquid aren't joined together as strongly as in a solid.

Liquids keep their volume but not their shape.

It won't squash. It feels like a solid

They keep their volume because the particles are held closely together. They change shape because the particles can slide over each other.

b Use the particle model to explain why the liquid will not squash.

Gases

How does the particle model fit what we know about gases? The pictures on the next page show what can happen with gases.

It feels like there's nothing there.

There is something there. I can feel my breath.

This balloon squashes.

C Look at the three pictures. Use the particle model to explain what is happening.

Using the model correctly

A 1 m rod of aluminium gets 2.3 mm longer if you heat it by 100 °C.

Yes, they just jiggle more so they take up more space.

So the particles stay the same size and are still touching?

That's easy – the particles get further apart.

That's not right. Don't forget that particles are always touching in a solid. If there were gaps between them, it would be a gas!

We can use the particle model to explain other observations, such as expansion.

Increasing the temperature of the aluminium rod increases the vibration of the particles in the solid. Each particle takes up slightly more space and so the solid **expands** – the rod gets longer.

Questions

Use the particle model to answer all these questions.

1 Look at the cartoon. The man is going to hurt himself when he hits the water.

The man would not hurt himself if he jumped like this onto a bouncy castle. Explain why he hurts himself hitting the water, but not when he hits the floor of the bouncy castle.

2 Look at the photo of a ship. Explain why the wind pushes the sails, making the boat move through the water.

Did you know?

At very, very high temperatures (100 000 °C and above), gases become 'plasma'. Matter becomes a plasma in extremely hot places, like inside the Sun. Plasmas give out light and other energy. The photo shows a scientist producing plasma in a laboratory.

For your notes:

- We can use the particle model to explain the behaviour of solids, liquids and gases.

- You have to be careful to use the particle model correctly.

- Solids and liquids **expand** because the particles vibrate more when they are heated.

Learn about:
● Diffusion
● Gas pressure

What's that smell?

The onion smells. How can we explain that using the particle model?

The smell is in the air, and the air is a gas. In a gas the particles are far apart and moving quickly. This means that the particles in a gas move and spread out to fill the space available. This spreading is called **diffusion**.

You cannot see what is making the smell in the air. The smell is caused by particles that are far too small to be seen. Special cells in your nose detect these particles. Particles from an onion can even get into your eyes and make them water.

Diffusion in action

If someone let off a stink bomb in the corner of the room, people sitting closest would be the first to smell it. Gradually people further and further away would smell it. Eventually everybody in the room would notice the smell.

The stink bomb contains a bad-smelling liquid. When you let the stink bomb off, this liquid evaporates and becomes a gas. The smelly gas particles mix with particles in the air. Particles of a gas move very quickly in all directions. The smelly particles keep moving at random and eventually **diffuse** through the room.

Diffusion happens because of the movement of the particles. It does not need any mixing or stirring.

ⓐ Why would a person sitting near the stink bomb smell it before someone sitting further away?

ⓑ Would a stink bomb spread quicker on a cold day or a hot day? Explain why.

Diffusion in liquids

Diffusion happens in liquids as well as gases. The photos below show diffusion happening in a liquid. In the first photo you can see purple crystals dissolving into the water. The purple colour then slowly diffuses, or spreads out. This is because the particles in a liquid slide over each other. The purple particles and the water particles mix until the purple particles are spread throughout the water.

ⓒ Will the particles in a solid mix by diffusion? Explain your answer.

Pump it up

Inside this balloon are millions of air particles. The gas particles are moving around in all directions.

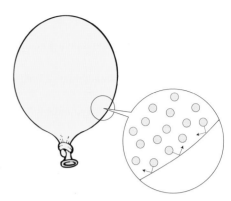

Quite often these particles bump into the sides of the balloon. Every time they hit the side of the balloon they give the rubber a tiny push, and transfer some energy to it. Each push is quite small, but lots of particles all pushing at the same time adds up to quite a lot of force. The force against the side of the balloon is the **gas pressure**. The balloon is kept in shape by the pressure of the gas inside it.

People often put balloons outside their houses to show that there is a party. During the evening, when the temperature drops, the balloons get smaller.

The gas inside the balloons has cooled. The particles have less energy and move more slowly. They hit the sides of the balloon less often and with less force. The sides of the balloon are pushed out less. The gas pressure is less. The balloon shrinks.

Air pressure

Air is a gas. The gas particles are all around us and we are always being hit by them. They cause air pressure (or atmospheric pressure). We don't usually feel the air pressure because the pressure within our bodies balances the pressure outside.

Wind is caused by air rushing from an area of high pressure to an area of low pressure.

Questions

1 Is diffusion faster in a gas or in a liquid? Explain your answer using ideas about particles.

2 Look back at the photos of the purple substance mixing with the water by diffusion. This diagram shows the particles at 12 o'clock. Draw another diagram to show the particles six hours later.

3 Dot noticed that the tyres on her bicycle were harder on a hot day than they were on a cold day. Explain why, using ideas about particles.

4 The afternoon of Peter's party was very hot. The balloons started to burst. Use your knowledge of particles to explain to Peter why the balloons burst.

For your notes:

- **Diffusion** is spreading and mixing caused by the movement of the particles.

- Diffusion happens in gases and liquids.

- Gases cause **gas pressure**.

- Gas pressure is because of the gas particles hitting a surface.

G5 Scientific models

Models

The word 'model' can mean lots of different things to us. You might think of a fashion model on a catwalk, or a model aeroplane, or you might model clay. When scientists talk about a **model**, they mean a way of showing how something works or looks that they cannot see or touch.

Christmas presents

On Christmas Eve, Jackie's six-year-old brother Matt had gone to bed. Jackie was looking at the presents for him under the Christmas tree to see if she could tell what they were. She couldn't open them, but she wanted to try and find out what was in them.

a **What could she do with the presents, apart from open them, that would help her collect more information about what they are like?**

Eventually, Jackie decided that one of the presents could be a book or a video.

b **Make a table to describe some of the ways a book and a video are similar, and some of the ways they are different.**

After shaking it, feeling it and testing its weight, Jackie decided the present was a video. She had collected her evidence and used her model of a video to explain what she could hear and feel inside the present.

Scientific models

Scientists use models in a similar way to help them think about how things work or behave. First they think about the evidence they have, and then they use a model that might explain it. You have met the particle model in this unit. Now try using it to explain the following four events.

1 Feeling the cold

This balloon was inflated with air and then dunked into liquid nitrogen (**A**). Liquid nitrogen is very cold indeed. When the balloon was taken out of the liquid nitrogen (**B**) it was a lot smaller.

c **Use the particle model to explain what is happening in the photos.**

d **Use the particle model to predict what will happen to the balloon as it warms up again.**

❷ Glass-blowing

The person in the photo is making a bottle by a process called 'glass-blowing'. A ball of glass is heated until it is red-hot. When it is hot, the glass can be stretched or shaped. This makes it easy to shape the glass into a bottle. The glass is then left to cool.

ⓔ What does this tell you about the particles in the glass when it is hot?

ⓕ Suggest what is different about the arrangement of glass particles in the cold glass bottle and in the hot ball of glass.

❸ Sublimation

This photo shows what happens when iodine crystals are heated gently over a Bunsen burner flame. Iodine does not form a liquid. It changes straight from a solid to a gas. Scientists call this **sublimation**. Crystals of iodine are forming again at the top the glass tube, where it is colder.

ⓖ Look carefully at the photo. Use the particle model to explain what is happening to the iodine:
(i) when it is heated   (ii) when it cools again.

❹ Breaking point

The photos below show a metal rod being tested to see how strong it is. The machine causes a very strong pulling force on the metal. The pulling force is gradually increased until the metal rod finally snaps.

ⓗ For each of the three photos, use your knowledge of the particle model to explain what is happening to the particles in the metal rod.

Questions

1 The polar ice-caps on the planet Mars are made from solid carbon dioxide. Carbon dioxide behaves in the same way as iodine. Use your knowledge of the particle model to explain why Mars would not flood if the planet warmed up.

2 Explain why scientists often need to use models to describe what is happening.

3 Glass panes in windows which are hundreds of years old get thicker at the bottom of the pane and thinner at the top. What does this tell you about glass?

4 Why may it be dangerous to add very hot water to very cold glass?

H1 Pure salt

Pass the salt

The clean, white substance we call salt does not start out that way. Most table salt is made from rock salt. Rock salt is dug out of the ground. It is a dirty, gritty mixture. You wouldn't want to put it on your chips!

Rock salt is a mixture of different substances. One of those substances is table salt, or **sodium chloride**. To get table salt, the sodium choride has to be separated from the other substances. We say that the sodium chloride is 'purified'. Sodium chloride on its own is a **pure** substance.

Do you remember?

When a solid dissolves, you can separate it by evaporating.

When a solid does not dissolve, you can separate it by filtering.

Purifying salt

Joe and Catherine are separating salt from rock salt.

There are all kinds of stuff in this. It's a real mixture. We'd never be able to pick out the salt by hand.

Salt dissolves in warm water. Maybe the other stuff won't.

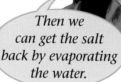

Good idea! Then we could filter it. The undissolved stuff will stay in the filter.

Then we can get the salt back by evaporating the water.

Joe and Catherine add warm water to the rock salt and stir. The amount of solid decreases as the salt dissolves.

a **Look at the photo of rock salt in water. How can you tell that it is a mixture rather than a solution of pure salt?**

Then they filter the mixture. The solid stays in the filter paper. The liquid that drips through is clear and colourless. It is salt **solution**. The salt has dissolved in the water because it is **soluble**. The sand and grit have stayed in the filter because they were **insoluble** in the water.

Joe and Catherine then pour the salt solution into an evaporating dish and heat the solution to get rid of about two-thirds of the water. They leave the rest of the solution to cool. The next day there are crystals of salt in the evaporating dish.

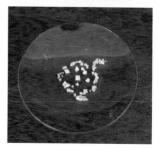

Reporting back

Joe and Catherine write about their experiment. Joe is keen to use the correct scientific words.

Catherine is more interested in using her ideas about particles. She draws diagrams to show pure water, the salt solution and pure salt.

Making a solution

The liquid that does the dissolving is called the **solvent**. Our solvent was water.

The solid that dissolves is called the **solute**. Our solute was the salt.

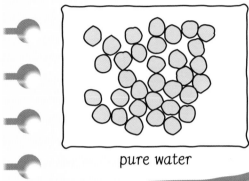

pure water

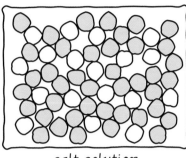

salt solution

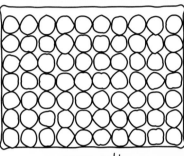

pure salt

b Which of Catherine's diagrams show a mixture?

Sue and Matthew wanted to know how much salt there was in the rock salt. They weighed the rock salt at the beginning and they weighed their salt crystals at the end. They then worked out the percentage of salt in the rock salt.

Sue thought that 70.5% of the rock salt was salt. Matthew said, 'No, *at least* 70.5% of the rock salt was salt. There may have been more.'

c Do you agree with Sue or with Matthew? Give your reasons.

Mass of rock salt: 20.0 g

Mass of pure salt: 14.1 g

Percentage of salt in rock salt $= \dfrac{14.1}{20.0} \times 100$

$= 70.5\%$

Questions

1 Use the glossary at the back of the book to explain these words containing the letters 'sol'. Present your work in an eye-catching way so that you will remember the meanings.

dissolve   insoluble   soluble   undissolved   solution   solvent   solute

2 How would you separate a mixture of powdered sulphur, iron filings (tiny pieces of iron) and salt? Use the information in the table to work out a method.

Substance	iron	sulphur	salt
Soluble in water?	no	no	yes
Magnetic?	yes	no	no

For your notes:

- **Insoluble** and **soluble** substances can be separated by dissolving and filtering.

- A soluble substance, or **solute**, dissolves in a **solvent** to make a **solution**.

- The solvent and the solute can be separated by evaporation of the solvent.

H2 Distillation

Getting drinking water

Imagine that you live on an island and the fresh water supply has been cut off. There is no fresh water for you to drink, only sea water. Sea water is a solution of salt in water. It is not good to drink. How could you separate the water and the salt to get drinking water?

If you heat the water, it will turn into a gas (water vapour) and **evaporate**. The salt would be left behind. To get drinking water, you can use a process called **distillation**. This process is shown in this diagram.

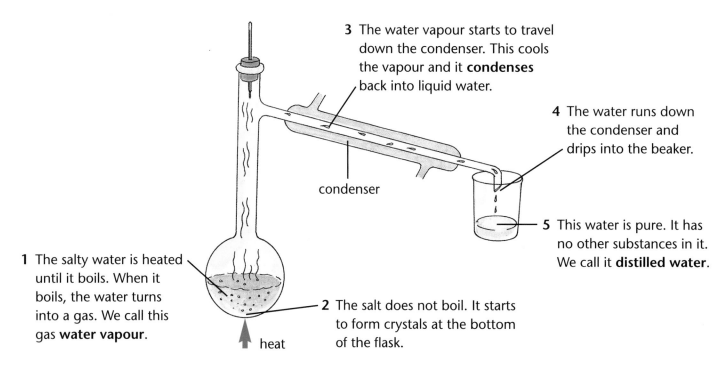

3 The water vapour starts to travel down the condenser. This cools the vapour and it **condenses** back into liquid water.

4 The water runs down the condenser and drips into the beaker.

condenser

5 This water is pure. It has no other substances in it. We call it **distilled water**.

1 The salty water is heated until it boils. When it boils, the water turns into a gas. We call this gas **water vapour**.

2 The salt does not boil. It starts to form crystals at the bottom of the flask.

heat

Separating ink

Joe and Catherine watch their teacher distilling red ink. The ink is a mixture of a red pigment dissolved in ethanol. The mixture is heated to 76°C, the boiling point of ethanol.

a Where are the following taking place:
(i) evaporation?
(ii) condensation?

Joe and Catherine are asked to draw diagrams showing the particles at **A**, **B** and **C**. Their diagrams are shown on the next page.

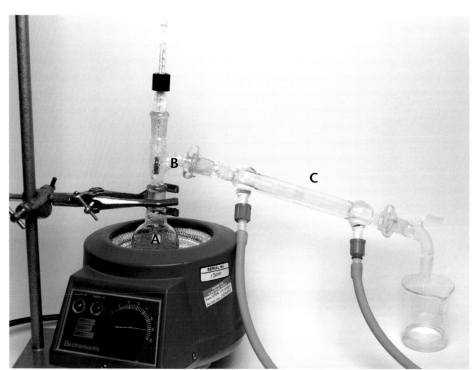

b Which diagram shows:
  (i) a solution?
  (ii) pure ethanol?

When all the ethanol has been evaporated, a red solid is left at A.

c Draw diagrams showing the particles at A when:
  (i) most of the ethanol has been evaporated
  (ii) all the ethanol has been evaporated.

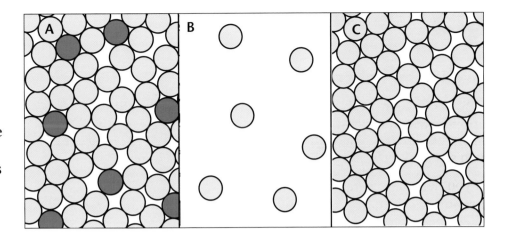

Industrial distillation

This photo shows a desalination plant. These are used in some countries to turn sea water into drinking water. They are quite expensive to run, because they need a lot of energy.

Mixtures of different liquids can also be separated by distillation. The mixture is heated. When the boiling point of the first liquid is reached, that liquid evaporates. The gas is captured, condensed and that liquid is separated. The mixture is then heated to the boiling point of the next liquid, which is evaporated and separated.

Crude oil is separated by in this way. The oil is heated to 400°C and almost all the substances in the oil evaporate. The mixture of gases is then cooled gradually and the different substances condense one by one. We make petrol from crude oil by distillation.

Questions

1 Put these statements in the correct order to describe distillation of salt solution to give pure water.

  A The solvent evaporates.

  B Pure solvent drips from the condenser and is collected.

  C The salt solution is heated until the mixture starts to boil.

  D The gas that enters the condenser is cooled and condenses.

2 Use your general knowledge and the information on these two pages to answer **a** to **c**.

  a What substances have boiling points of: **(i)** 100°C? **(ii)** 76°C?

  b If you heated a mixture of water and ethanol, which substance would boil first?

  c Name two mixtures separated on a large scale by industry.

For your notes:

- Substances with different boiling points can be separated by **distillation**.

- Distillation involves heating to **evaporate** and cooling to **condense**.

- Distillation can be used to separate a pure liquid from a solution.

H3 Chromatography

Separating inks

The ink in your pen is probably not made of one colour. It is a mixture of colours or **dyes**. To separate them out, we can use a method called **chromatography**.

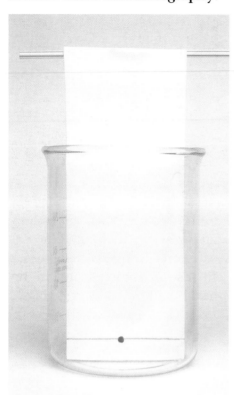

Did you know?

Scientists have improved chromatography so it can be used to separate hundreds of different mixtures, not just inks and dyes.

1 The mixture is spotted onto the filter paper. The pencil line shows where the mixture was put at first.

2 The solvent must start below the ink spot. The solvent travels up the paper. The different coloured inks travel at different speeds.

Analysing foods

Many foods have artificial colours added to make the colours brighter. Chromatography can also be used to look at these colourings.

The Pluto Sweetie firm brought out a new range of orange and yellow sweets called 'Brighties'. Tony was worried that the sweets might contain a food colouring called sunburst yellow. He is allergic to sunburst yellow but not to other yellow colours.

Tony used chromatography to separate out the colours in the yellow Brightie and the orange Brightie. He compared these with four different yellow food colours: sunburst yellow, solar yellow, mellow yellow and sunny yellow.

The black line and the crosses show where the ink mixtures were before the separation.

ⓐ Which type of sweet should Tony avoid? Give your reasons.

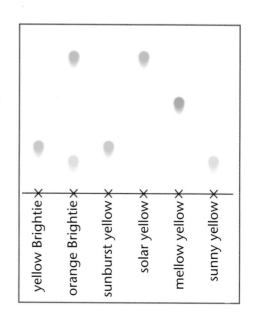

82

How does it work?

How far each dye moves depends on how soluble it is. Substances that are very soluble in the solvent move a long way up the paper. Substances that are not very soluble in the solvent only move a short distance up the paper.

Look at the diagram showing particles. The yellow particles are the paper. The pale blue particles are the solvent. You are looking at the paper from the side.

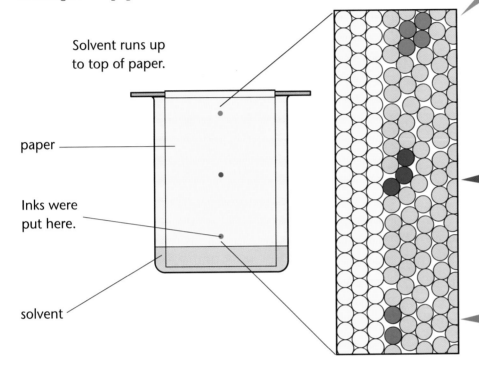

Solvent runs up to top of paper.

paper

Inks were put here.

solvent

The green dye is very soluble in the solvent. There are forces between the green particles and the particles of solvent. There are no forces between the green particles and the particles of paper. The green dye travels with the solvent up the paper.

The purple dye is slightly soluble in the solvent. There are forces between the purple particles and the particles of paper _and_ forces between the purple particles and the solvent. The purple dye particles spend some time stuck to the paper, and some time travelling with the solvent. The purple dye moves up the paper, but not as quickly as the solvent.

The red dye is insoluble in the solvent. There are forces between the red particles and the particles in the paper, but none between the red dye and the particles of solvent. The red dye stays where it was put, stuck to the paper.

ⓑ **Put the three dyes in order of solubility, with the most soluble first.**

Questions

1 Look at the separation of inks **A**, **B**, **C** and **D** on the right.

   a How many different dyes are present?

   b Which is the most soluble?

   c Which is the least soluble?

   d Are any of the dyes totally insoluble in the solvent? Explain your answer.

2 Look back at the separation of the yellow and orange food colours.

   a Rank the four food colours according to:

     (i) solubility in the solvent (most soluble first)

     (ii) pulling forces between the particles of dye and the paper (strongest forces first).

   b What do food companies normally do to help people find out what additives are in their foods?

3 Is the ink in your pen a mixture? Plan an experiment to find out.

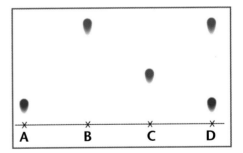

A   B   C   D

For your notes:

- **Chromatography** is used to separate a mixture of **dyes**.

- The dyes are separated because they are more or less soluble in the solvent.

- Chromatography can be used to identify dyes by comparing them to named dyes in the same experiment.

83

Changing cheques?

Shaheen is a forensic scientist. It is her job to help the police solve crimes. She uses different methods to examine clues. It is important that she analyses the results of her investigations very carefully

Shaheen's latest case is about a cheque for £1999. Mr Jones says that he wrote the cheque for £1000. Mr Crisp says that the cheque was always for £1999.

If Mr Jones is telling the truth, then the words 'nine hundred and ninety nine' were added by Mr Crisp after Mr Jones signed the cheque.

If Mr Crisp is telling the truth, then all parts of the cheque were written at the same time with the same ink.

Lyons Bank
CAMBRIDGE BRANCH

90-27-13

Date 5 May 03

Peter Crisp

One thousand nine hundred and ninety nine   £1999

ALBERT JONES

Albert Jones

"000632"   90-2713   2927163

One ink or two?

Shaheen decides to analyse the cheque using chromatography. First she photographs the cheque. Then she cuts out the words 'one thousand' and the words 'ninety nine'. She dissolves the ink using a solvent, filters the mixture to remove the lumps of paper, then evaporates the solvent.

Shaheen uses chromatography to separate the dyes in the two inks. First she used the solvent ethanol. Then she does the separation again using the solvent propanone. The pages from her laboratory record are shown below. Look at Shaheen's comments under the heading 'Analysis'.

a Why does Shaheen think that the bright blue dye is insoluble in ethanol but very soluble in propanone?

b Why does Shaheen write that there are 'at least four coloured dyes' after the first experiment but 'at least five coloured dyes' after the second experiment?

c Why do the same inks give two different patterns?

d Do you think the two inks are the same? Give reasons for your answer.

e Do you think the cheque was always for £1999? Give reasons for your answer.

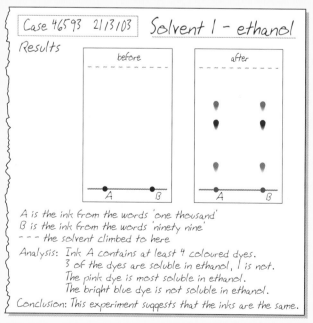

Case 46593 21/3/03   Solvent 1 – ethanol
Results

before   after

A   B   A   B

A is the ink from the words 'one thousand'
B is the ink from the words 'ninety nine'
- - - the solvent climbed to here
Analysis: Ink A contains at least 4 coloured dyes.
3 of the dyes are soluble in ethanol, 1 is not.
The pink dye is most soluble in ethanol.
The bright blue dye is not soluble in ethanol.

Conclusion: This experiment suggests that the inks are the same.

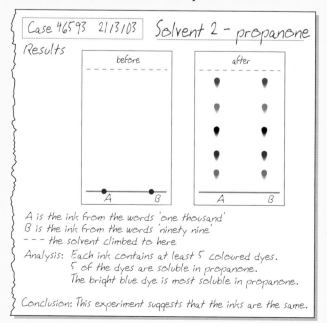

Case 46593 21/3/03   Solvent 2 – propanone
Results

before   after

A   B   A   B

A is the ink from the words 'one thousand'
B is the ink from the words 'ninety nine'
- - - the solvent climbed to here
Analysis: Each ink contains at least 5 coloured dyes.
5 of the dyes are soluble in propanone.
The bright blue dye is most soluble in propanone.

Conclusion: This experiment suggests that the inks are the same.

A different light

Shaheen decides to look at both 'after' chromatograms using UV light. The page from her laboratory record is shown on the right.

f What extra information is visible using UV light?

g In your opinion, are ink A and ink B the same? Give your reasons.

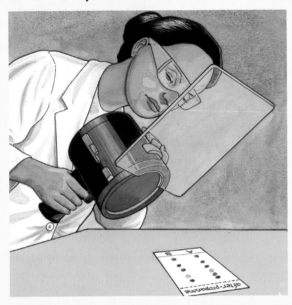

Case 46593 21/3/03 Viewing with UV light
Results

after – ethanol   after –propanone

A   B   A   B

A is the ink from the words 'one thousand'
B is the ink from the words 'ninety nine'
— — — the solvent climbed to here
Analysis: Ink A contains at least 6 dyes, 5 visible under
   normal light and 1 visible under UV light
   Ink B contains at least 5 dyes, all visible under
   normal light. There is no evidence of a dye that
   is visible under UV light in ink B
Conclusion: This experiment suggests that the inks are different.

In her conclusion, Shaheen does not say that the inks are definitely different. She says that the experiment suggests that the inks are different.

h Can you think of different reasons why there might be an extra dye in ink A but not in ink B? Describe and explain your ideas.

i How would looking at the cheque with UV light help Shaheen confirm her results?

j How would repeating the chromatography confirm her results?

k If the inks were identical, would it prove that Mr Crisp is innocent?

Questions

1 Do you have enough evidence to say how many dyes are in ink **A** and in ink **B**? Explain your answer.

2 During the trial, Mr Crisp's lawyer suggests that Mr Jones may have changed pens half way through writing the cheque.

  a Suggest a reason why Mr Jones may have changed pens.

  b Do you think this is likely? Give your reasons.

3 Imagine you are Shaheen presenting her evidence to the jury at the trial.

  a Write an account of what you would say to the jury.

  b Would you include the results from the ethanol experiment in your explanation? Explain how you came to your decision.

  c How would you answer these questions from Mr Crisp's lawyer:

    (i) 'How certain are you that the inks are different?'

    (ii) 'Is there any evidence that the words were written at different times?'

Why is it soluble?

The photo on the right shows copper sulphate dissolving in water. The diagram shows what is happening to the particles.

In solid copper sulphate, there are forces holding the copper sulphate particles together.

In solution, the copper sulphate particles are surrounded by water particles.

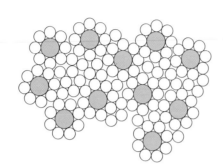

During dissolving, water particles surround the copper sulphate particles. The forces holding the copper sulphate particles together are replaced by forces between the water particles and the copper sulphate particles.

Where does it go?

Look at the three photos. They show 5 g of salt and 100 g of water being weighed. When the salt is dissolved in the water the solution weighs 105 g.

Matter is still there when you make a solution. You have the same mass of solute and solvent as when you started. We say that the mass has been **conserved**. You have not lost any particles, so you do not lose any mass.

Saturation and solubility

You can add more and more copper sulphate crystals until no more will dissolve. The solution is then **saturated**.

Scientists measure what mass of a substance will dissolve in 100 cm^3 of solvent at a certain temperature. This is called the **solubility** of the substance. Different substances have different solubilities.

Different temperatures

The solubility of a substance depends on the temperature of the solvent. The graph on the next page shows how much of substances X and Y will dissolve in 100 cm^3 of water at different temperatures.

Look at this graph.

a How much of substance X dissolves at 70°C?

b How much of substance Y dissolves at 10°C?

c Which substance is the more soluble at 60°C?

Temperature also affects how quickly a substance will dissolve. At higher temperatures, the solvent particles have more energy and move more quickly. The solvent particles hit the lumps of solid more often and harder, knocking off the outer particles more easily.

Cooling off

When a saturated solution cools, the dissolving process is reversed. Crystals begin to grow and the solute comes out of solution.

Look again at the graph.

d How much of substance X dissolves at
(i) 80°C?   (ii) 55°C?

e How much of substance X would crystallise out if you cooled a saturated solution from 80°C to 55°C?

Different solvents

Different solutes dissolve in different solvents. For example, salt dissolves in water but not in petrol. For a substance to dissolve, there must be pulling forces between the solute particles and the solvent molecules. Common solvents include water, ethanol and propanone. Propanone is often used as nail varnish remover.

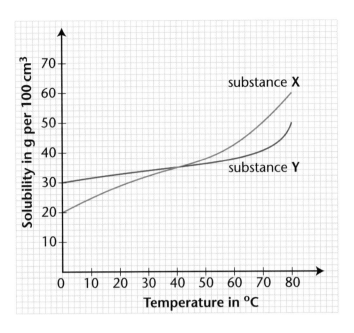

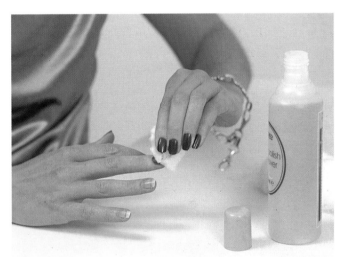

Questions

1 What is meant by 'solubility'?

2 Use the solubility table to answer these questions.

Substance	Solubility in g per 100 cm³ of water at 25°C
magnesium sulphate	43
sodium chloride	36
copper sulphate	22
magnesium chloride	53

a How much magnesium chloride will dissolve in 500 cm³ of water?

b Can you make a saturated solution by adding 10 g of copper sulphate to 50 cm³ of water? Explain your answer.

For your notes:

● **Solubility** is the mass of a substance that will dissolve in 100 cm³ of solvent.

● The solubility of a substance changes with temperature.

● Solutes dissolve more quickly at higher temperatures.

● Mass is always **conserved** during dissolving.

● A substance can be soluble in one solvent and insoluble in another.

I1 Energy on the move

Energy resources

Talking about energy

In science, we use the word 'energy' in a special way. **Energy** makes things happen.

That boy has lots of energy.

You were very energetic today.

Eat your breakfast. It will give you energy.

I haven't the energy to do that!

On the move

If something is moving, it has energy. We call this **movement energy** or **kinetic energy**. The athlete in the photo is running. He has lots of kinetic energy.

Sounding out

When a jet plane goes very fast, you hear a booming sound. When there is a loud sound, there is a lot of energy. If there is less energy the sound is softer.

If something makes sound, it is giving out energy. We call this **sound energy**.

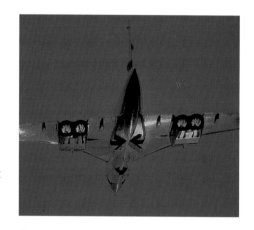

Lighting up

If something is giving out light, then it is giving out energy. We call this **light energy**. The electric light in the photo is giving out light energy. It is a **light source**. A light source gives out light energy. The most important light source is the Sun.

Hotting up

When something is hot, it is giving out energy. The metal in the photo is very hot. It is giving out lots of energy. We call this **thermal energy** or **heat energy**.

a How do our bodies detect: **(i)** light energy?
(ii) sound energy? **(iii)** thermal (heat) energy?

Plug it in

Look at the picture of the iron. It only works if it is plugged in and turned on. It needs energy from electricity to work. We call this energy **electrical energy**. Electrical energy comes from power stations or from batteries.

Look at the photo of the electricity meter. It measures the amount of electrical energy coming into the house from the power station.

ⓑ **What other things on this page need electrical energy to work?**

Moving energy around

Energy moves from place to place. Look at the drawing of the lamp. The energy comes into the lamp as electrical energy. It comes out of the lamp as light energy and thermal energy. We say that the energy is **transferred**.

We show **energy transfers** using arrows. We write on the arrows to show how the energy is coming in and going out. This is an **energy transfer diagram**.

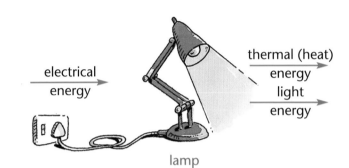

lamp

Energy makes things happen

Heat, light and sound energy are all energy. We talk about light energy and sound energy, but there aren't really different types of energy. It is just easier to say 'light energy' rather than 'energy that our bodies detect with our eyes'.

You can think about energy being like money. Money buys things, like energy makes things happen. You can pay with coins or notes, by cheque or by credit card, but it is all money!

Questions

1 Look at pictures **A** to **D**. Which picture shows something:

  a with lots of kinetic energy?

  b giving out light energy?

  c giving out sound energy?

  d giving out thermal (heat) energy?

  e working because of electrical energy?

2 Copy and complete this energy transfer diagram.

  kettle

3 Draw and label your own energy transfer diagram for:

  a an electric cooker   **b** a TV.

For your notes:

● **Energy** makes things happen. Things work because of energy.

● We sense **light energy**, **sound energy** and **thermal (heat) energy**.

● Things that move have **kinetic (movement) energy**.

● Electricity carries energy. We call this **electrical energy**.

● When energy is moved about, we say it is **transferred**.

Taking the strain

Look at the picture of the archer. Where was the energy before it was moving the arrow? It was stored in the stretched bow and bowstring. We call energy that is stored in something stretched **strain energy**.

Look at the picture of a toy. It moves because it has been wound up. Winding the toy stores energy in a spring inside the toy.

Fuelled up

Look at the picture of a bonfire. The fire is giving out light energy and thermal (heat) energy. Where was the energy before the bonfire started? The energy was stored in the wood. We call energy stored in materials **chemical energy**.

Wood is a **fuel**. All fuels are stores of chemical energy. They give out energy when they are burned.

Food is also a store of chemical energy. Our bodies release the energy stored in the food.

Another store of chemical energy is **batteries**. Batteries use the stored chemical energy to give out electrical energy.

Lifted up

Look at this picture. The bucket and water have kinetic (movement) energy. Where was the energy before the water and the bucket started moving?

The energy was stored in the bucket and the water because they were lifted up. Things that are lifted up have energy because of **gravitational attraction** (gravity). We call energy stored because of gravity, **gravitational energy**.

a Think about a jet plane. What types of stored energy does it have?

Showing stored energy

We can show stored energy in a box. The diagram shows the stored energy in a bonfire. It is shown in the box. The energy given out when the wood is burned is shown by the arrows.

chemical energy in wood	→ thermal energy
	→ light energy
	→ sound energy

CLASS 7B

In and out of storage

Look at the girl on the trampoline. The girl's weight stretches the trampoline. The trampoline pushes the girl into the air.

b When is the most energy stored in the trampoline? How is it stored?

c When is the most energy stored in the girl as gravitational energy?

d What happens to the energy when it is not being stored as strain energy or gravitational energy?

Energy can be transferred in and out of storage.

Talking more about energy

On page 89, money was used as a model for energy. Money can be transferred using coins, notes, cheques or credit card. We can take the model further.

Money can be stored. It can be stored in a wallet or a purse or a safe. It can be stored in a bank account or a building society. When money is spent, it is like energy being transferred. When money is saved, it is like energy being stored.

Questions

1 How is the energy stored in **a** to **f**?

   a a stretched rubber band   **b** a snowball at the top of a hill

   c a chocolate bar   **d** a skydiver jumping out of a plane

   e a firework   **f** a squashed ball.

2 Draw and label an energy transfer diagram for the bow and arrow shown in the picture at the top of the opposite page.

3 Draw and label an energy transfer diagram for the girl and trampoline. Make sure you include all the stages shown in photos **A** to **C**.

4 Describe the money model for energy. Do you think it is a useful way of thinking about energy?

For your notes:

- Energy stored because a material is being pulled or pushed is called **strain energy**.

- Energy stored in fuels, food or **batteries** is called **chemical energy**.

- Energy stored in an object because it is lifted up is called **gravitational energy**.

- **Fuels**, food and batteries are all stores of **chemical energy**.

I3 Energy in food

Measuring energy

Energy is measured in **joules**, symbol **J**. One joule is the energy needed to lift an apple (weight one newton) by one metre.

A joule is very small. One joule is the amount of energy needed to increase the temperature of only $1\,cm^3$ of water by about one quarter of a degree Celsius. This means that a bath of hot water contains hundreds of millions of joules. The numbers are too big to handle, so we use **kilojoules** instead. The abbreviation for kilojoule is **kJ**.

> 1 kilojoule = 1000 joules

One kilojoule:

- will raise the temperature of $10\,cm^3$ of water by almost 24 °C
- is the kinetic energy of a girl running at 6 metres per second
- is the energy given out by a bright light bulb in 10 seconds
- is the energy needed to play music very loud for 40 seconds.

Energy from food

Without food we would starve and die. Food is the fuel for our bodies. Our bodies use oxygen to release the energy stored in our food. Food is a store of chemical energy. Our bodies use the energy that is released from food for everything we do, such as moving, and also to keep us warm. So we transfer chemical energy to kinetic energy and thermal energy.

We get fat if we eat too much food and take too little exercise because energy is stored in the body. We get thin if we eat too little food. The amount of energy we need from food depends on what we do.

The bar chart shows how much energy you need to do different tasks for one hour.

a How many kilojoules do you use to:
   (i) sleep for one hour?
   (ii) walk uphill for 15 minutes?

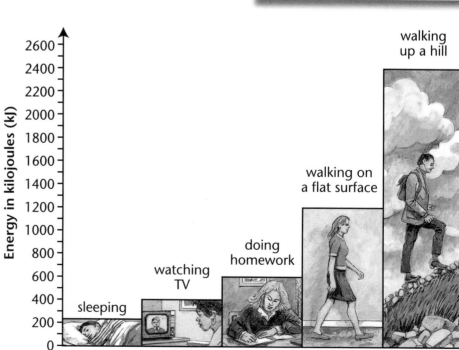

Energy in kilojoules (kJ)

sleeping — watching TV — doing homework — walking on a flat surface — walking up a hill

Telling how much energy is in food

Processed foods come with a nutritional information label. This label tells us about the food. One of the things it tells us is the amount of energy in the food.

b **How many hours could you watch TV using the energy in half a chicken and leek pie?**

Nutritional information from a chicken and leek pie

Average values	per 100 g	per half pie
Energy	**1050 kJ**	**1975 kJ**

Unprocessed foods, like fruit, vegetables and fresh meat, do not come with a nutritional information label. You have to look up the energy content of foods like these in tables which tell you the energy for every 100 g of each food.

Food	Energy per 100 g	Mass of one portion
cornflakes	1550 kJ	30 g
milk	268 kJ	100 g (for cereal) 24 g (in coffee)
bread	1006 kJ	35 g
butter	2970 kJ	10 g
marmalade	1108 kJ	15 g
grapefruit	326 kJ	150 g

c **Frank is on a diet. He wants to have a breakfast that uses 1500 kJ of his daily allowance. Make up a breakfast for Frank using the information in the table. (*Hint:* you will need to work out the energy in one portion of each food first.)**

Where does the energy come from?

Our food is made up of plants and animals. Plants get their energy directly from the Sun. They trap energy from the Sun to make their food. Light energy is transferred to the plant and the plant then stores some of that energy as chemical energy.

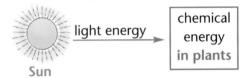

All animals get their energy from plants. Either they eat plants, or they eat animals that eat plants. This means that animals, including humans, also get their energy from the Sun.

Questions

1 Explain how the energy in a fried egg comes from the Sun. The egg was fried in butter.

2 Imagine that you are a doctor writing for a magazine. Write an article to advise teenagers how to lose weight sensibly.

3 Think back to the money model on pages 89 and 91. Do joules and kilojoules fit into the model? Explain your answer.

For your notes:

● Energy is measured in **joules, J** and **kilojoules, kJ**.
1 kilojoule = 1000 joules.

● Food contains energy that originally came from the Sun.

● We take in chemical energy in our food and give out energy as kinetic energy and thermal energy. Any energy not given out is stored in the body.

14 The best fuel

Investigating fuels

Shaibal's class investigated the energy given out by two different fuels, lighter fuel and firelighters. The class tested each fuel in turn.

They burned the fuels and heated water with them. The fuel that heated up the water more gave out more energy.

a The class decided to use the same mass of each fuel. Why do you think they did this?

b They decided to heat the same volume of water with each fuel. Why do you think they did this?

The pictures below show Shaibal and Pippa's experiment with lighter fuel.

1 They used 100 cm³ of water and 2 g of fuel.

2 They took the temperature of the water at the start and found that it was 21 °C.

They then did another experiment using firelighters instead of lighter fuel. They wanted to compare the firelighters with the lighter fuel.

Shaibal and Pippa burned 5 g of firelighter and heated 100 cm³ of water. The temperature of the water was 21 °C at the start and 82 °C at the end.

3 They took the temperature when all the fuel had burned away and found that it was 46 °C.

4 They took the start temperature away from the end temperature to find out how much the fuel had heated the water:

```
  46
- 21
  25 °C
```

c By how much did the fuel heat the water?

d Do you think this was a fair test? Explain your answer.

e Would you have done the experiment in the same way as Shaibal and Pippa? Explain your answer.

Variables

There are three things that could be different at the start of this investigation. They are **variables**. They are:

* the type of fuel
* the mass of fuel
* the volume of water.

The class are investigating the type of fuel, so this is the only one they change. It is called the **input variable**.

The change in water temperature depends on the type of fuel being used. This is called the **outcome variable**. This is the variable that they measure.

Lighter fuel again

Shaibal and Pippa then did a different investigation. They used lighter fuel this time. Their results are shown in the table below.

Mass of fuel in g	Amount of water in cm³	Temperature of water at start in °C	Temperature of water at end in °C	Temperature rise in °C
1.0	100	21	33	12
1.5	100	21	38	17
2.0	100	21	46	25
2.5	100	21	52	31
3.0	100	21	59	38

f What was the input variable for this investigation?

g What was the outcome variable?

h What variables did Shaibal and Pippa keep the same to make it a fair test?

i Make a line graph of Shaibal and Pippa's results. Put the input variable along the bottom and the outcome variable up the side. Draw the best straight line you can through the points.

We use the word **relationship** to describe how the outcome variable changes when the input variable is changed.

j Is there a relationship between the temperature rise (the outcome variable) and the mass of fuel (the input variable)? If so, describe it.

The temperature rises less as you burn a bigger mass of fuel.

There is no relationship between the rise in temperature and the mass of fuel you burn.

The temperature rises more as you burn a bigger mass of fuel.

15 Fossil fuels

What are fossil fuels?

We find **coal**, **oil** and **natural gas** inside rocks. All three are fuels because when they are burned, they release lots of energy. They are called **fossil fuels** because they formed from animals and plants that lived millions of years ago. A **fossil** is the dead remains of an animal or plant preserved in rock.

How did fossil fuels form?

Coal is made from plants that lived millions of years ago. They grew using energy from sunlight (1). When they died, the plants were buried (2). The dead plants did not rot in the usual way. This was because rotting needs oxygen, and the dead plants were buried away from the air.

Instead, the plants changed, slowly, into coal. This change was caused by the weight of many layers of mud and sediment on top of the plants (3). As the plants turned into coal, the layers of mud and sediment turned into rock (4). These changes took many millions of years.

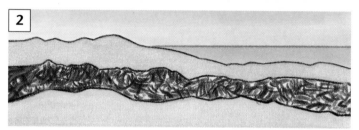

a Write a 'recipe' for making coal.

Oil and natural gas formed in a similar way, but they were made mostly from animals. Many millions of years ago, like now, the sea was full of tiny animals. These animals lived by eating plants, which made their food by trapping the energy in sunlight.

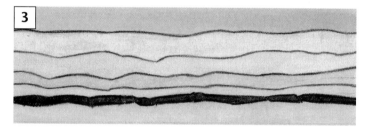

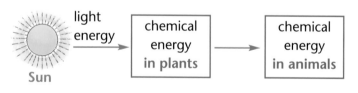

Sun — light energy → chemical energy in plants → chemical energy in animals

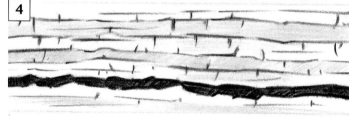

When these animals died they sank to the bottom of the sea where there was very little oxygen. Without oxygen, the tiny sea animals did not rot. Instead, they were buried by layers of sediment forming above them. The weight of these layers changed the dead sea animals into crude oil and natural gas, while the sediments around them changed into rock. Again, these changes took many millions of years.

b Make a series of labelled drawings showing the formation of crude oil and natural gas.

Using fossil fuels

We burn fossil fuels in our homes to heat water and to keep us warm. We make petrol, diesel and other fuels from crude oil. These are burned in cars and other vehicles. However, most of our fossil fuels are burned in power stations to make electricity.

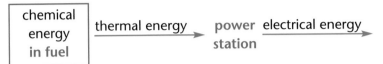

We also use fossil fuels, particularly oil, to make useful materials like plastics.

Non-renewable energy resources

Fossil fuels are going to run out because they are **non-renewable energy resources**. This means that we are using them up far faster than they can be replaced. Remember, they took many millions of years to form.

It is difficult to decide how soon we will run out of fossil fuels. It depends on how fast we use them, and it depends on whether we find more oil fields and coal deposits. Scientists suggest we will run out of oil in about 2030, natural gas in about 2050 and coal in about 2230.

At the moment, 90% of our energy comes from fossil fuels. We need to conserve our reserves of fossil fuels, so that they last longer. We also need to find other ways to get the energy we need. We need to find some alternative energy resources.

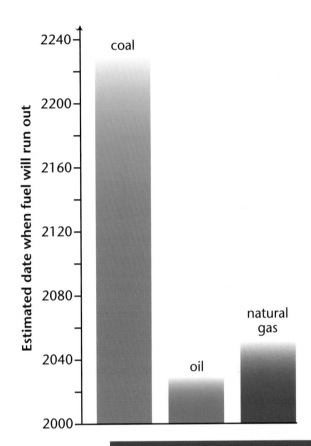

Questions

1 Which of statements **A** to **D** are true and which are false?

A The energy stored in oil came from the Sun.

B Fossil fuels could be renewed within ten thousand years.

C Oxygen is needed for fossil fuels to form.

D Coal was made mainly from dead plants.

2 **a** How many years before we use up all the:

(i) coal? (ii) natural gas? (iii) oil?

b We cannot be sure about the answers to question **2a**. Why not?

3 Imagine life in 100 years' time when all the natural gas and oil have run out and coal is rare and expensive. List all the ways everyday life could be different.

For your notes:

Fossil fuels:

● include **coal**, **oil** and **natural gas**

● were made from dead animals and plants that lived many millions of years ago

● contain energy that came from the Sun

● provide 90% of our energy needs, including our electricity

● are **non-renewable energy resources** that are running out.

Using fuels wisely

Making it last

Fossil fuels are non-renewable. They will run out. They will last longer if we use them more carefully. This is called **conserving** fossil fuels.

Everyone needs to stop wasting energy. We need to share cars rather than driving everywhere alone. We need to drive cars with smaller engines. We need to insulate our homes properly, so less thermal energy escapes. That will mean we burn less fuels heating our homes. We need to turn off lights to save energy. Then we use less electricity and that means the power stations burn less fuel.

We can also save fuel by recycling more materials. It takes less energy to recycle glass than to make new glass. The same is true for metals like aluminium and copper, and for some plastics.

a **Explain how the following will help us conserve energy:**
   (i) **sharing cars**
   (ii) **driving cars with smaller engines** **(iii) insulating our homes**
   (iv) switching off lights **(v) recycling glass.**

SAVE ENERGY!

Switch off	Turn off lights when you leave a room. Turn off TVs, computers, and radios when you're not using them – many constantly use small amounts of power.
Light efficiency	Use energy efficient light bulbs wherever possible. They use less electricity and last up to six times longer.
Go green	Buy your home's electricity from a renewable energy source. See www.greenprices.com for more information.
Recycle	80-90% of household rubbish ends up in landfill sites. Recycling saves landfill space, energy, raw materials and cuts air pollution.

- Use mains electricity wherever possible – or rechargeable batteries (they can be recharged up to 1000 times).
- Recycling glass and plastics conserves energy and oil reserves. 70% of plastics could be recycled – but only 1% currently is.
- Recycling aluminium cans instead of using raw materials can cut the energy needed by 90%.

Biomass is renewable

We need to find a replacement for fossil fuels before they run out. Fossil fuels were made from plants and animals, so a good place to start is with the plants and animals that are alive today. Plant and animal material is called **biomass**. Biomass contains chemical energy and can be used as fuel. instead of fossil fuels. It is an **alternative energy resource**.

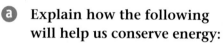

Sun → light energy → chemical energy in biomass → thermal energy → power station → electrical energy →

Wood has been used by humans as a fuel for many thousands of years. We could plant trees and harvest them, like a crop. Some trees, like pine, grow quickly. It would take 10 to 15 years to grow a crop of wood. However, wood is still a **renewable energy resource** because we can replace it.

Wood can be used to make **charcoal**. This is done by baking the wood. Some of the energy stored in the wood is lost, but the charcoal made burns at a much higher temperature than 'raw' wood. Nowadays we use charcoal on our barbecues because the high temperatures cook the food well. Charcoal was used in the past to melt metals like iron.

People from other cultures use other forms of biomass as fuel.

The Inuit people (Eskimos) live in the far north. They burn blubber as a fuel. Blubber is the layer of fat found under the skin of seals.

The Bedouin people live in the desert. They use camel dung as fuel.

ⓑ **Explain why neither the Inuits nor the Bedouin use wood as a fuel.**

Rotting material makes **methane**. Natural gas is mainly methane, so we could use methane from rotting material as a replacement for natural gas. Any plant or animal material could be rotted to make methane. The tank in the photo on the left contains farm waste which is used to make methane.

ⓒ **Why is it a good idea to use rubbish and waste to make methane?**

Should we burn biomass?

The problem with all fuels is that they produce waste gases when they burn. Biomass fuels make carbon dioxide and water when they burn. The water is harmless, but the carbon dioxide may be dangerous for the environment. Scientists think that increasing amounts of carbon dioxide in the air are causing global warming. This will make sea levels rise and weather patterns change, destroying many habitats (including those of humans).

Hydrogen

Another alternative fuel is hydrogen. Hydrogen makes only water when it burns, so it burns very cleanly. Unfortunately, it is not easy to handle. It is a gas so it takes up lots of space. It is very lightweight and floats away. Also, a mixture of hydrogen and oxygen can explode.

ⓓ **Think about using hydrogen as a fuel. What are:
(i) the good points? (ii) the bad points?**

Questions

1 Where does the energy stored in biomass come from? Explain your answer.

2 a Compare wood with coal. Why is wood renewable and coal non-renewable?

  b Why is it important to plant wood crops rather than cutting down established forest and woodland?

  c What are the advantages of charcoal as a fuel?

3 a Make a list of all the biomass fuels mentioned on these two pages.

  b Why do we need to start using more of these fuels?

  c We used to use whale oil as a fuel. Why did we stop?

4 Explain how biomass can be used to make electricity.

For your notes:

- We need to **conserve** fossil fuels so that they last longer.

- **Biomass** is plant and animal material. It is a store of chemical energy.

- Wood and **methane** are examples of biomass that can be used as fuel.

- Many forms of biomass are **renewable energy resources** because the plants and animals grow quickly.

Solar energy

Solar furnaces are huge, curved mirrors that reflect all the sunlight to a single point. The thermal energy can then be used to make electricity in a power station.

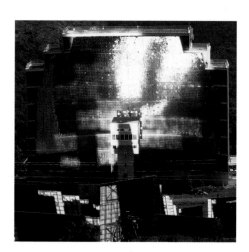

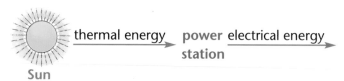

Sun — thermal energy → power station — electrical energy →

Solar energy can also be used to heat water in homes, using solar panels. Water flows in pipes through the solar panel, where it is heated by the Sun. The hot water then flows to the hot-water tank ready for baths or washing up.

Solar energy is renewable because the Sun shines every day. However, solar furnaces are very expensive to build and only work well in full sunshine.

a Why do we not build solar furnaces in Britain?

Wind energy

We can use the wind to make electricity. Look at the **wind turbine** in the photo on the left. There is a **generator**, like a tiny power station, in the box at the top. It takes in kinetic energy from the wind, or **wind energy**, and makes electricity.

Winds are caused by the Sun heating the air. The Sun heats the atmosphere every day, so wind energy is a renewable energy resource. Wind turbines can be noisy and some people think they are ugly. They also only work when it is windy!

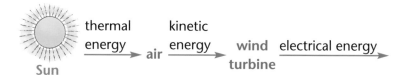

Sun — thermal energy → air — kinetic energy → wind turbine — electrical energy →

b Where would you build a wind turbine?

Wave energy

Wave energy can also be used to make electricity by turning a **wave turbine**. Waves are made by wind passing over the water. As wind is a renewable energy resource, so are waves. However, wave turbines are still experimental and not very reliable.

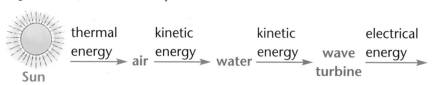

Sun — thermal energy → air — kinetic energy → water — kinetic energy → wave turbine — electrical energy →

Falling water

We use falling water to generate electricity. This is called **hydroelectric power (HEP)**. The water needs to be high up, so it is a store of gravitational energy. When the water falls it has kinetic energy. A **water turbine** takes in the kinetic energy and produces electrical energy.

The water has to be high up before it can fall through a turbine. If the hydroelectric power station is high in the mountains, then the water was lifted by the Sun.

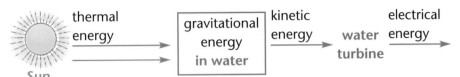

Geothermal energy

The Earth's core is hot. This is called **geothermal energy**. In a few places there are hot rocks close to the surface. These can be used to heat homes. In a very few places the rocks are hot enough to boil water. In these places geothermal energy is used to make electricity. Geothermal energy does not come from our Sun. The Earth's core is still hot from when the Earth was made, billions of years ago.

C Do you think geothermal energy is renewable?

Less burning for cleaner air

Burning fossil fuels, like burning biomass, is increasing the levels of carbon dioxide in the air, causing global warming. Burning coal and oil also releases sulphur dioxide, which causes another pollution problem called **acid rain**. Solar energy, wind energy, wave energy and falling water do not involve burning, so do not contribute to air pollution.

Questions

1 Copy and complete this table.

Key:		
R = renewable	**NP** = non-polluting (does not involve burning)	
S = energy from the Sun		
H = heats homes	**HEP** = hydroelectric power	

Energy resource	R	S	H	NP	Disadvantages
fossil fuels					
biomass fuels					
solar energy					
wind energy					
wave energy					
HEP (mountains)					
geothermal					

For your notes:

- Alternative energy resources include **solar energy, wind energy** and **wave energy**.

- These are all renewable energy resources because they get their energy from the Sun.

- **Hydroelectric power** from falling water is a renewable energy resource that gets its energy from the Sun.

- Energy resources that do not involve burning do not contribute to air pollution.

J1 Electrical energy

Electricity everywhere

You know that electricity can light our homes and streets, warm our houses and make our washing machines, TVs, microwave cookers and computers work. But how does it do all this?

In the diagram the **battery** is making the lamp light. The correct word for a single 'battery' is a **cell**. More than one cell is called a battery. A cell provides energy and the electricity is what carries energy to make things work and to make the lamp light.

The diagram shows the energy transfer that happens in the circuit on the right.

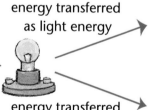

energy stored as chemical energy in the cell

energy transferred as electrical energy

energy transferred as light energy

energy transferred as thermal energy

The cell stores **chemical energy** which is transferred to the circuit as **electrical energy**. The electrical energy is transferred to the lamp as thermal energy we can feel and light energy we can see.

ⓐ **What transfers energy to the circuit?**

ⓑ **Draw an energy transfer diagram for a portable stereo.**

Voltage tells us how much energy the electricity is carrying. The bigger the number of **volts**, the more energy the cell stores. The short way for writing volts is **V**. The number of volts is called the voltage of the cell.

The energy must be transferred along the wires to the lamp.

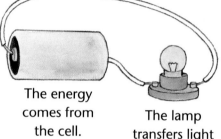

The energy comes from the cell.

The lamp transfers light energy and thermal energy.

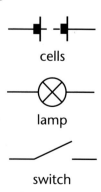

Some of the batteries we buy are made up of cells. A nine volt battery, like the one in the picture on the right, contains six 1.5 V cells.

Used cells have lost most of their stored energy so they won't make things work.

At the end of the 1700s, an Italian count called Alessandro Volta invented the first battery. Each cell in his battery had two metal discs in a salt solution. Volta got the best results with copper and zinc and the current increased if he piled up alternate discs of the metals separated by paper soaked in salt solution. This powerful battery was called 'Volta's pile'. The volt is named after Volta.

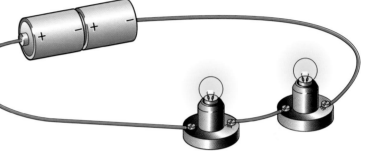

The modern 'torch battery' was invented by Leclanche. It has a carbon rod instead of copper and the zinc is the outer case.

Positive to negative

If you are connecting cells in a circuit to make a battery, make sure you connect the positive end of each cell to the negative end of the next one.

Questions

1   Inderjit found a selection of cells in the cupboard. They were marked with different voltages, 1.5 V, 4.5 V and 9 V. Which one

  a   stores the most energy?

  b   is capable of lighting most lamps?

2   Look at circuit **Y**. All the switches are shown open.

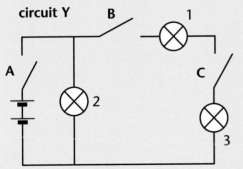

circuit Y

Switches closed	Lamps lit
A, B, C	1, 2, 3
none	
A, C	
A, B	
A	

  a   How many lamps are there in this circuit?

  b   How many switches are there in this circuit?

  c   Copy and complete the table.

3   Describe Volta's invention and how it led to the development of the 'modern battery'.

For your notes:

- We get **energy** from electricity to make things work.

- You need a **complete circuit** for energy to be transferred.

- **Cells** store energy. More than one cell connected together is called a **battery**.

- Cells with a high **voltage** store more energy.

Current affairs

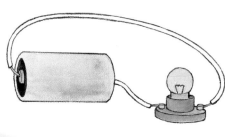

Current

You need a complete circuit to transfer energy from the battery to the lamp. We are going to learn about what is happening in the circuit. Electricity flows in the wires. We call this a **current**. The current flows around the whole circuit.

The electrical energy in this circuit is transferred as heat and light energy. It moves out of the circuit into the air and has to be replaced by more energy from the cell.

It helps to think about it like a central heating system where the water in the system is given heat energy by the boiler. The flow of water carries the heat energy to the radiators in the house. Then the heat energy is transferred to the radiators and into the air.

Just like the current in an electrical circuit is not used up, the water is the central heating system is not used up. There is always the same amount of water in the system.

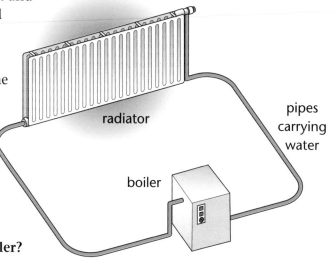

radiator

pipes carrying water

boiler

a What part of the electrical circuit is like the boiler?

b What part of the circuit is like the radiator?

Measuring current

You can show that current is never used up by measuring it in different places in the circuit.

You can show the current in the wires by putting an **ammeter** in the circuit. The circuit symbol for an ammeter is Ⓐ. An ammeter measures current in **amps**. The short way of writing amps is **A**.

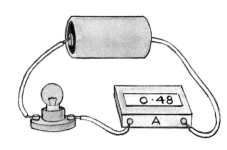

Moving the ammeter

Ben, Laura and Dan were doing an experiment about current. Their teacher asked them where to put the ammeter in a circuit.

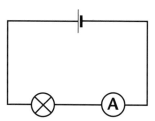

Does it matter where you put the ammeter?

It will have to be in the circuit. Let's put it to the left of the lamp.

Ben

It doesn't matter where you put it. The current is the same on both sides of the lamp.

Laura

No, energy leaves at the lamp. I think the current will be different on either side. Let's use two ammeters to check.

Dan

Their experiment is shown here.

c Is the current different or the same on each side of the lamp?

d Who was correct, Ben, Laura or Dan?

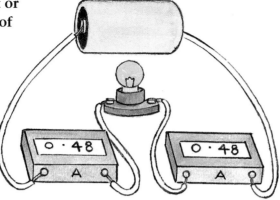

More complicated circuits

If all else is the same, the more the current the brighter the bulb.

Ben and Laura decided to investigate this question:

● If you have more cells, does the current change?

The circuits they used and their results are shown below.

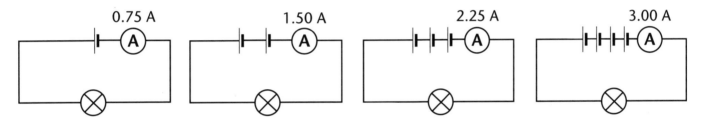

e What happens to the current as the number of cells increases?

If you put more cells in the same circuit, you get more energy.

Adding more cells to a circuit is like putting a more powerful pump in your central heating system. The bigger the pump, the faster the water will flow round the system.

> ## Do you remember?
>
> Adding more cells to a circuit makes the lamp brighter.

Questions

1 Draw a circuit diagram with a cell, a lamp and an ammeter.

2 What unit is current measured in?

3 **a** Use Ben and Laura's results to plot a graph of current against number of cells. Put the number of cells along the bottom and the current up the side. Draw the line of best fit. It should be a straight line.

b Ben and Laura did not use five or six cells. Suggest why they stopped at four cells.

4 Do you think the central heating model is a good one for explaining current. Think about what happens if a radiator leaks, compare this with breaking a circuit.

For your notes:

● **Current** flows around a circuit. It is measured in **amps**, **A**, using an **ammeter**.

● The current is the same on both sides of a lamp.

● Current is not used up as it flows around a circuit.

● Adding more cells increases the current in a circuit.

Flowing through

If you think about water flowing through pipes, it flows more easily through wide pipes than through narrow ones. Narrow pipes slow the flow down.

Thin and thick wires in a circuit are like narrow and wide pipes. A thin wire slows the current down because it is harder for the electricity to get through. The thin wire resists the electricity more so we say it has a high **resistance**. Resistance makes it hard for the current to flow.

Components like light bulbs slow the current down. Because the current flows around a whole circuit, it slows down all around the circuit.

Using resistance

An ordinary light bulb has a very thin wire (the **filament**) inside it and this has a very high resistance. It is this resistance which causes the transfer of energy in an electrical circuit.

Humphrey Davy first made light using electricity in 1801 but it took nearly 80 years to find a thin filament which would glow and give out light without burning away. Eventually, in 1878 and 1879 two inventors, Joseph Swan and Thomas Edison, almost at the same time came up with the answer – a carbon filament sealed in a glass bulb with no air inside. This meant that the carbon glowed rather than burned.

In 1911 the carbon filament was replaced by a tungsten filament, which lasted longer. Then in 1913 the filaments were coiled, so there was more wire to glow.

Thomas Edison.

Joseph Swan.

Dimmer and dimmer...

Laura and Dan made a dimmer switch out of a pencil lead. They set up a circuit like the one shown in the diagram on the right. As they moved the crocodile clips along to make the pencil lead longer, the lamp became dimmer. This is because it was harder for the current to flow through the longer pencil lead.

a What happens to the resistance as you change the length of the pencil lead ?

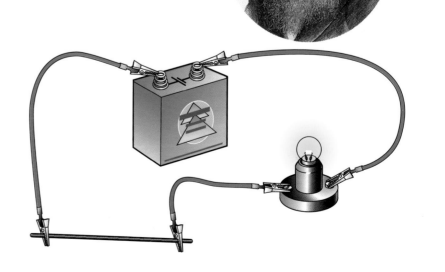

Series circuits

In **series circuits** the lamps are arranged side by side in the same loop as shown here. The more lamps there are, the more the current is slowed down through the whole circuit.

A lower current collects energy less often from the battery and so less energy gets to the lamps. This means that they shine less brightly than just one lamp in the circuit.

b What current is flowing through the lamps in the series circuit?

series circuit

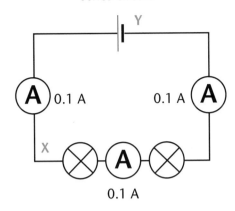

Parallel circuits

There is a way of connecting several lamps to the same size cell but keeping them as bright as just one would be. You can put them in different loops as shown on the right. This is called a **parallel circuit**.

c What current is flowing through the lamps in the parallel circuit?

d Which lamps are brighter, those in the series circuit or the parallel circuit?

e Look at the series circuit. What would the current be at X and at Y?

f Look at the parallel circuit. What would the current be at P, at Q and at R?

parallel circuit

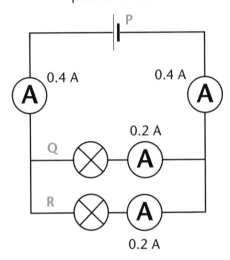

In the series circuit, the current is the same at all points in the circuit. In the parallel circuit the current is shared between the loops of the circuit.

Questions

1 Explain the meaning of the word resistance. Describe how resistance is useful in dimmer switches.

2 What would happen to the current in a series circuit if a lamp with a higher resistance were used?

3 Make a table with the headings 'Series circuits' and 'Parallel circuits'. Fill in the table to compare:
   ● lamp brightness
   ● number of loops
   ● current.

4 Use the information of these pages to make a time line called 'The invention and development of the electric filament lamp'.

For your notes:

● **Resistance** makes it hard for the current to flow.

● You can connect lamps in **series** and in **parallel**. **Parallel circuits** have more than one loop.

● Two lamps in parallel are brighter than the same two lamps in series, using the same cells.

● In a **series circuit** the current is the same at all points.

● In a parallel circuit the current is shared between the loops.

J4 Models of electricity

Using models

Scientists use **models** to help them think. Each part of a model represents something in real life.

A good model fits with the facts. So far, you know these facts about electricity.

Electricity carries energy to make things work.

You need a complete circuit to transfer energy from a cell to make a lamp light.

The current is the same on both sides of the lamp.

Resistance makes it hard for the current to flow.

We have used models of a central heating system and water flowing through taps to talk about current in a circuit. We can use other models for electrical circuits.

The coal truck model

This model shows a mine and a power station. There is a single-track railway between the mine and the power station. Coal trucks run along the railway. The coal trucks can move quickly or slowly along the track.

At the mine, the coal trucks are filled with coal. The coal trucks run along the tracks and deliver the coal to the power station. The empty coal trucks then return to the mine.

Read the description carefully again and study the diagram. Then answer the questions.

mine

power station

ⓐ In the coal truck model, what represents:

   (i)   the circuit?   (ii) the cell?
   (iii) the lamp?   (iv) the energy?

ⓑ The moving trucks represent the current. When the trucks speed up, does the current increase or decrease?

ⓒ If part of the track was narrower than the rest, what would this represent?

ⓓ Do you think this is a good model? Explain your answer.

Models are very useful to explain how things work but the problem with a model is that it is never perfect.

In the coal truck model, if the coal trucks are derailed, coal will spill out onto the tracks.

ⓔ What happens if there is a break in an electrical circuit? Does electricity spill out?

The class and matches model

Mrs Fuller is using another model to explain electricity to her class.

Mrs Fuller gives each pupil a match as they pass her.

The pupils continue and collect another match.

The pupils carry their matches round the white circle.

Mrs Huxley strikes each match as the pupil passes it to her.

e Draw a diagram of this model. Use the same colours as in the coal truck model:

- energy is green
- the circuit is pink
- the current is yellow
- the voltage (where energy goes into the circuit) is blue.

f Does this model help you understand electricity? Explain your answer.

g Can you put an ammeter into your class and matches model? What would the ammeter do? Where would the ammeter be in your drawing?

h How could you show resistance using this model?

Questions

1 Which was the better model for you, the coal truck model or the class and matches model? Explain your answer.

2 Can you think of another model for electricity? Draw a diagram to explain your model. (Other people have suggested a roller coaster or a bicycle.)

J5 Electrical hazards

Finding a fault

A **fault** is something that stops an electrical circuit from working. You can find a fault in a circuit by replacing each part of the circuit until it works again. If your torch didn't work you would first replace the battery to see if it worked again.

You can also find a fault by testing each part of the circuit to see if it is working properly. The best way to test a component is to put it in a circuit that you know is working.

Circuits in the home

The lamps in the rooms of a house are connected in parallel. This is because if you switch off the light in one room, the others are not affected. Look at the diagram, if you switch off the lamp in the kitchen, the lounge lamp stays on. If the lamps were arranged in series, switching off one would switch off all the others.

In homes this type of circuit is called a **ring main**. One advantage of a ring main is that when one bulb does not work, all the others keep on working.

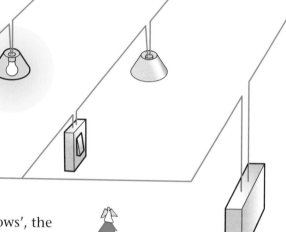

Some sets of fairy lights are connected in series. If one lamp 'blows', the circuit is broken and all of the others go out. The advantage of using a series circuit is that a lower current flows through the lights and is less dangerous.

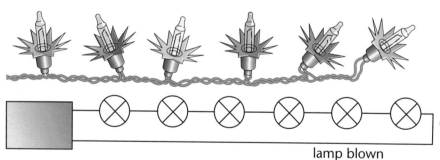

lamp blown

Why can electricity be dangerous?

Electricity from the mains supply delivers much more energy than the energy we get from the cells we use in radios, clock, and cd-players. This is why we should be very careful when we use mains electricity at home.

If the current gets too high or there is a fault in the circuit, a **fuse** can stop current flowing. This can stop you getting an electric shock or stop the wires getting hot and causing a fire.

Do you remember?

If you get an electric shock from mains electricity (240 V), it can stop your heart beating.

Fuses work because fuse wires have a high resistance. As the current increases, it has difficulty flowing through the fuse so energy is transferred as light and heat. If the current gets too large, the heat causes the fuse wire to melt and break the circuit.

Electricity in your body

Your nerves use very small electric currents to send signals to all parts of your body. Your heart beats because of a regular jolt of electricity through it. But just as you need these very small electric currents for your body to work properly, a much larger current could kill you. People have died when their kites or fishing rods have touched the overhead power lines.

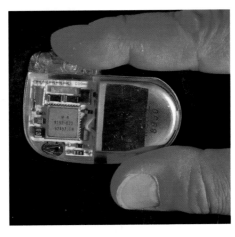

A heart pacemaker sends a small pulse of electricity to the heart to keep it beating.

In the late 1700s an Italian anatomist called Galvani showed the effect of an electric current on frog's legs. He noticed that the legs of the frogs he dissected twitched when electrical sparks were produced by a nearby machine. He was puzzled because it only happened when the frog's legs were hung on a metal hook. He thought that electricity caused muscles to move and that there was a special kind of electricity in living things.

But later Volta realised that it was the metal hook that was important and went on to invent his batteries using metals and salt solution.

Questions

1 Jane has a torch that does not work. Write down all the faults that could stop the torch working. How could Jane check for each of the faults?

2 You have two lamps and three switches. You want to make a parallel circuit that will:
   ● switch off both lamps together
   and also
   ● switch off each lamp separately.
   Draw the circuit that will do this.

3 Suggest two reasons why a parallel circuit is used to connect the lamps in a house.

4 What happens to the current in a ring main as more lamps are switched on at night?

For your notes:

● We can find a **fault** by testing each part of the circuit.

● The kind of circuit we use in our homes is called a **ring main**.

● Electricity can be dangerous. It can stop your heart beating.

● **Fuses** work by breaking the circuit if the current is too high.

K1 Forces and gravity

Forces everywhere

Everything you do uses forces. You cannot see forces, but you can often see the effects of a force. A force can change the shape of an object, or make it move faster or slower, or make it change direction. The greater the force, the greater its effect. 'Push' and 'pull' are two types of forces. The magnetic attraction between a magnet and iron is another type of force. All forces are measured in **newtons** or **N**.

What is gravitational attraction?

Gravitational attraction (gravity) is the force which pulls an object towards the Earth. The picture shows the Earth is shaped like a sphere. Britain and Australia are almost on opposite sides.

Gravitational attraction pulls Sharon and Shirley toward the Earth. The force acts down towards the centre of the Earth.

a **Why does Shirley not fall off Australia?**

There is also gravitational attraction between the Moon and objects on it. The gravitational attraction on the Moon is weaker than on the Earth.

Do you remember?

You have used a forcemeter (newtonmeter) to measure forces. The 'newton' is the unit for measuring force.

Sharon

Shirley

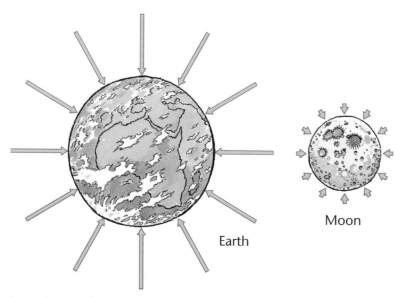

Earth

Moon

What is weight?

When you pick an object up off the floor, you are lifting it against a force. This force is the object's **weight**, which pulls it down. In the picture, Samson has to pull with a force greater than the dumbbell's weight to pick up the dumbbell.

We use 'weight' to mean how heavy something is. Weight is gravity acting on an object. Heavy objects are pulled down with a bigger force than light objects. We measure weight in newtons.

Samson's pull

dumbell's weight

Mass and weight

Sharon is made of a certain amount of stuff or **matter**. Sharon's **mass** is a measure of how much matter she is made of. Mass is measured in **kilograms** or **kg**. She has a mass of 66 kg.

Sharon's weight is different. Her weight is caused by gravitational attraction on her mass. On Earth gravitational attraction has a force of 10 N on each kilogram. So, to find the weight of something on Earth, you multiply its mass by 10. On Earth a mass of 1 kg has a weight of 10 N.

Sharon's mass is 66 kg

Shirley's mass is 45 kg

b What are the weights on Earth of Sharon and Shirley?

On the Moon

The force of gravitational attraction between two objects depends on the mass of the objects. The Moon has a smaller mass than the Earth, so the Moon has less gravity than the Earth. In fact it is one-sixth of the Earth's. On Earth, Sharon weighs about 660 N. On the Moon she weighs only 110 N.

weight 660 N
mass 66 kg

weight 110 N
mass 66 kg

c Explain why Shirley would weigh less on the Moon than on Earth.

Questions

1 Explain what these words mean:

   a weight   b mass   c gravitational attraction.

2 Calculate the weight on Earth of these people:

   a Susan, mass 70 kg   b Philippa, mass 55 kg

   c Marco, mass 88 kg.

3 Sharon's rabbit has a mass of 6 kg. What is the weight of the rabbit on:

   a the Earth?   b the Moon?

4 Explain how mass affects the gravitational force.

For your notes:

- **Gravitational attraction** is the force that pulls an object and the Earth towards each other.

- **Weight** is the force of **gravitational attraction** on an object. We measure weight in **newtons, N**.

- **Mass** is a measure of how much **matter** an object is made of. We measure mass in **kilograms, kg**.

K2 Friction

What is friction?

Friction is a force that is exerted when things rub together. Friction can slow things down. The ice skates have very little friction on the smooth ice, so you can skate fast. The runner's shoes have good grip and make lots of friction with the ground to help her slow down and stop.

a  What kind of surfaces make the most friction?

b  What kind of surfaces make the least friction?

Friction can be useful

Friction can be a very useful force. Bikes and cars have brakes that use friction to slow them down or stop them. The surfaces of the brakes rub against the wheels so the wheels don't turn so fast. Some road junctions have a special high-friction surface to slow cars down in case they skid as they stop.

c Give some more examples showing how friction is useful in everyday life.

Reducing friction

Sometimes friction is not useful and we want to reduce it. When two surfaces rub together, they will eventually become worn down because of friction.

Machines have a lot of parts that rub together. To reduce friction, we use substances such as oil and grease. We call these **lubricants**. They make surfaces run smoothly against each other. Surfaces that are smooth or are separated by air also have little friction.

Engine oil

Engine oil is a lubricant. It reduces friction and helps the pistons to move easily. In the winter, the engine oil gets very cold and thick so it doesn't flow as easily. Engines do not run as smoothly until the oil has warmed up. An engine must start easily on a freezing cold day and keep going on a long journey at temperatures as high as 90 °C.

Did you know?

A hovercraft rides on a cushion of air. This reduces friction between the hovercraft and the water.

Friction makes things warm

Where there is friction, heat energy is given out. You can feel this happen when you rub your hands together. They feel warm.

Did you know?

It is illegal to drive a car with worn down treads or 'bald' tyres because friction between the tyre tread and the road prevents skidding on wet roads. But some racing cars have smooth tyres. This is because the rubber becomes very hot at high speed and sticks to the road!

Elijah McCoy

Elijah McCoy invented the 'Real McCoy Lubricating Cup'. Not many people know of his work. He was born in Ontario, Canada in 1844. Elijah's parents escaped from slavery in the USA. He studied mechanical engineering at Edinburgh University in Scotland and then returned to work in America.

Elijah worked on the railroads oiling the engines of trains. In his spare time he developed a self-lubricating cup that sent oil automatically to the engine, so that the train did not have to stop to be oiled. Elijah's device was described as the 'real McCoy'. It became so popular that people would ask if new equipment contained the 'real McCoy'. We still use the expression 'real McCoy' to mean the 'real thing'

d **Explain why is engine oil is less effective as a lubricant in cold weather.**

Elijah McCoy and his invention.

Questions

1 **a** Describe how friction can be useful.
  b Describe how the effects of friction can be reduced.
2 Write a story about a world without friction.
3 How do we know that heat energy is given out when there is friction?
4 Describe the problem that Elijah McCoy solved and his solution.
5 Plan a fair test to measure the thickness of engine oil at different temperatures. You are provided with water baths at different temperatures, thermometers, funnels, conical flasks, engine oil, measuring cylinders and stopclocks.

For your notes:

- **Friction** is a force that is exerted when things rub against each other.

- We can reduce or increase friction to make it useful to us.

- We use **lubricants** to reduce friction between moving parts.

Staying put

Look at the picture of Zena and Sam having a tug-of-war. They are not moving. They are pulling with the same sized force, but in opposite directions. The forces are shown with **force arrows**. A force arrow points in the direction of the force. The length of the arrow shows the size of the force. If two forces are the same size and pull in opposite directions, the forces are **balanced**.

Do you remember?

An object stays where it is because the forces are balanced.

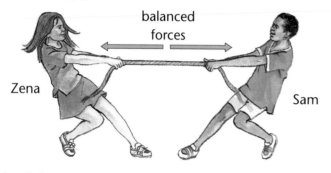

balanced forces

Zena   Sam

a Which of the arrows A, B and C shows the biggest force?

b Which arrow shows the smallest force?

A   B   C

More balanced forces

The picture shows a mass hanging from a spring on a forcemeter. The mass is not moving. The spring has stretched. The amount it stretches is called the **extension**. The extension depends on how much mass is hung on the spring. The forces on it are balanced. Weight is the force pulling down on the spring with the same force as the spring pulls up on the mass.

If the weight is too heavy, the spring will be stretched out of shape. The spring does not return to its original shape when you take the mass off.

Mr Blue the decorator stands on a plank of wood to paint a wall. He stands very still. The plank bends down because of Mr Blue's weight. The plank pushes up. This force from the plank is called a **reaction force**. The force pushing down on the plank is the same as the force pushing up on Mr Blue. The forces are balanced. If they were not, the plank would break.

force of spring

weight

c What are the forces on your chair when you sit still on it? Draw a diagram with arrows.

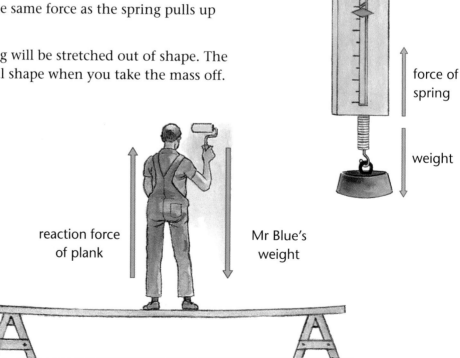

reaction force of plank

Mr Blue's weight

d Look at the diagrams below. They all show situations where the forces are balanced on a non-moving object. Sketch each diagram and add the force arrows.

Why do things float and sink?

When you put an object into water, the water pushes up on the object. This force is called **upthrust**. If the object **floats**, then the upthrust is equal to weight. The forces are balanced. If the object **sinks**, then the upthrust and weight are not equal. The forces are unbalanced.

e Draw a force diagram to show the forces on a floating boat.

There are balanced forces on a hot-air balloon floating in air.

Steady speed

When you are travelling in a car at a steady speed, the forward forces are the same as the forces of friction acting against the car. The picture below shows this. The forces are balanced when the car is moving at a steady speed. So balanced forces don't just exist when an object is still. They can also exist when an object is moving at a steady speed.

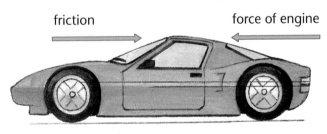

friction   force of engine

Questions

1 Copy the sentences below and say whether you think the forces are balanced for each one.

   a A car travelling at a steady speed on a motorway…

   b A car speeding up to overtake a slow lorry…

   c A car parked at the roadside…

2 Name the force that balances your weight when you stand on a plank.

3 Julie and Jill pull on a rope with a joint force of 50 N in one direction, and Jack and Paulo pull in the opposite direction. If the forces are balanced, what force do Jack and Paulo exert?

4 Draw a force diagram showing the forces on an aeroplane travelling at a steady speed.

For your notes:

- If two forces are the same size and pull in opposite directions, they are called **balanced forces**.

- The **reaction force** stops something falling through a solid object. The reaction force balances the weight.

- When an object **floats**, the forces of weight and **upthrust** are equal. If the forces are unbalanced, the object **sinks**.

- When a car is moving at steady speed the forces are balanced.

Unbalanced forces

The box in the picture has two forces acting on it. One is the pull of the rope. The other is the weight of the box. The force arrows are different lengths. This tells us that the forces are different sizes. They are called **unbalanced forces**.

a Which force is bigger?

b In which direction do you think the box will move?

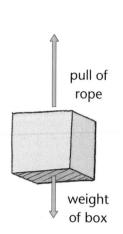

pull of rope

weight of box

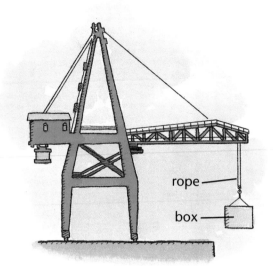

rope

box

Getting going

If Dipal does not push his go-kart, it will not start moving. A force is needed to start something moving.

Dipal gave his go-kart a gentle push. It did not move at all.

c What force do you think stopped the go-kart moving?

Dipal gave his go-kart a bigger push. Dipal's push on the go-kart was bigger than friction, so the go-kart started to move.

When forces push against each other like this, and one force is bigger than the other, they are unbalanced forces.

When there are unbalanced forces acting on an object, the object starts to move. It moves in the direction of the bigger force and it gets faster.

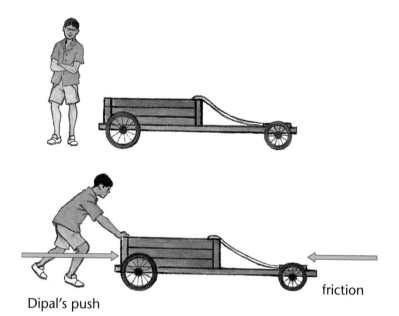

Dipal's push

friction

Calculating the size of the force

In this diagram, the forces are unbalanced. The big man will push the box towards the small man. We can work out the size of the force that pushed the box:

The size of the unbalanced force is sometimes called the **resultant force**. In this example the resultant force is 75 N.

25 N

100 N

$$100\,N - 25\,N = 75\,N$$

Shaping up

Unbalanced forces can also change the shape of an object. It might become bent, twisted or even break.

d Give an example of something which changes shape with unbalanced forces on it.

Unbalanced forces on moving objects

Unbalanced forces can act on something that is already moving. The car in the diagram is moving forwards.

force of engine   air resistance

Air makes friction with moving objects such as cars and planes. We call this **air resistance**. Because the force from the engine is bigger than air resistance, the car moves faster.

e The force from the engine is 1000 N, the force from air resistance is 50 N. What is the resultant force?

When the bigger force is in the same direction as the movement, the object speeds up.

When the bigger force is in the opposite direction to the movement, the object slows down.

Unbalanced forces on moving objects can also make them change direction. If you are pushing a luggage trolley in a straight line and someone walks into it from the side, the trolley will change direction and move sideways.

Do you remember?

Air resistance is a force that slows down objects moving through air. Try walking into the wind!

Questions

1 What happens when there are unbalanced forces acting on an object that is not moving?

2 What happens to a moving object when there are unbalanced forces acting on it and the bigger force is in the same direction as the movement?

3 What happens to a moving object when there are unbalanced forces acting on it and the bigger force is in the opposite direction to the movement?

4 What might happen to a foam cushion when there are unbalanced forces acting on it?

5 Helen pushed a sledge with force of 8 N. The force of friction pushed against this with a force of 2 N. What is the resultant force?

For your notes:

- **Unbalanced forces** can act on an object that is not moving. The object starts to move in the direction of the bigger force.

- The size of the unbalanced force is called the **resultant force**.

- Unbalanced forces can also change the shape of an object.

- Unbalanced forces on a moving object can make the object speed up, slow down, or change direction.

K5 Slow down!

Talking about speed

People use different sayings to describe how fast or slow things move.

We can tell how fast a thing moves by measuring its **speed**.

How do we measure speed?

To find the speed of an object, you need to know the distance the object travels and the time it takes to travel that distance. You take the distance travelled and divide by the time taken. We can show it like this:

speed = $\dfrac{\text{distance travelled (in metres, m)}}{\text{time taken (in seconds, s)}}$

We measure distance in metres or kilometres and time in seconds or hours. In science, we measure speed in **metres per second** or **m/s**. In everyday life, we find it easier to measure speed in **kilometres per hour** or **km/h**.

As fast as the speed of light.

As swift as a deer.

Slower than a snail.

Faster than a speeding bullet.

Calculating speed

Example A dog ran 30 metres in 2 seconds.

dog's speed $= \dfrac{30\,\text{metres travelled}}{2\,\text{seconds taken}}$

$= \dfrac{30\,\text{m}}{2\,\text{s}}$

$= 15\,\text{m/s}$

The dog ran at a speed of 15 metres per second or 15 m/s. This means it ran 15 metres every second.

ⓐ **Find the speed of these people.**
● **Danny walked 20 kilometres in 4 hours.**
● **Susan ran 100 metres in 20 seconds.**
● **Yin travelled 200 kilometres in 2 hours on a train.**

Distance/time graphs

You can draw a distance/time graph to show whether someone is moving fast or slowly on a journey. The graph shows Moira's journey home from school.

ⓑ **How far is it from Moira's school to her home?**

ⓒ **How long did she wait at the bus stop?**

ⓓ **When was Moira travelling fastest? How could you tell this from the graph?**

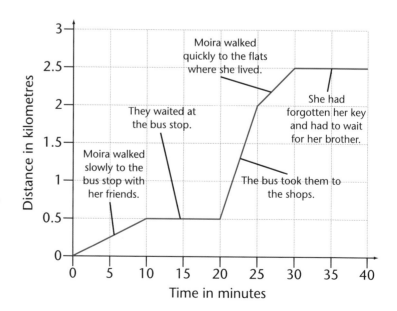

Moira walked quickly to the flats where she lived.

They waited at the bus stop.

She had forgotten her key and had to wait for her brother.

Moira walked slowly to the bus stop with her friends.

The bus took them to the shops.

Stopping distances

When a driver decides to stop, it takes a fraction of a second to step on the brakes. The car travels several metres in this time. This is called the **thinking distance**.

After the driver puts a foot on the brake the car slows down but continues to travel further until it stops. This is called the **braking distance**. The overall **stopping distance** is the thinking distance + the braking distance.

The faster a car is travelling, the longer it takes to stop and the further it travels while stopping.

30 mph

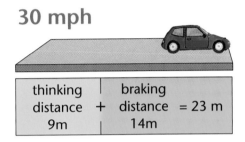

| thinking distance 9m | + | braking distance 14m | = 23 m |

60 mph

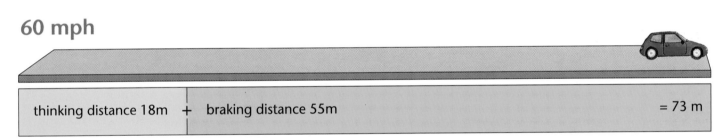

| thinking distance 18m | + | braking distance 55m | = 73 m |

The picture shows stopping distances for a car with good tyres in good weather. The tread on the tyres will grip the road and there will be more friction to slow down the car. A car will take longer to stop if the tyres are worn, or the road is wet, because there is less friction.

Questions

1 Think up some new sayings such as 'faster than a speeding bullet' for the following words:

   a speed   **b** fast   **c** slow.

2 How do we work out the overall stopping distance for a car?

3 What is the relationship between speed and stopping distance?

4 A car travels 120 kilometres in 2 hours. What was its average speed?

5 Describe two factors that can affect the stopping distance of a car.

For your notes:

- **Speed** is the distance an object travels in a certain time.

- The units used for speed are **metres per second, m/s,** or **kilometres per hour, km/h.**

- You can draw a distance/time graph to show whether something is moving fast or slowly on a journey.

- The faster a car is travelling, the longer it takes to stop.

K6 Archimedes' story

A new image

King Hiero reigned in ancient Greece. One day he decided that it was time for a new look.

'I know what I need – a brand new crown,' he thought. He sent for the court goldsmith and gave him a lump of gold to make the crown.

Crowning glory

Days later, the goldsmith delivered the crown. 'You look a real treat, your Highness,' he said. The next day, the King began to have doubts. There were rumours about the honesty of the goldsmith. The King suspected that the goldsmith had kept some of the valuable gold and put cheap silver in the crown.

ⓐ **What would you do if you had suspicion like the King's?**

Archimedes to the rescue

Hiero sent for his adviser, the brilliant young scientist Archimedes. 'I am sure you are the person to prove that the goldsmith is a thief,' he said. 'Just be sure you don't damage the crown.'

Back home, Archimedes realised that solving this problem would be hard if he could not damage the crown. He decided to have a bath. Still lost in thought, he took off his clothes and climbed into the bath, not noticing that it was full to the brim! As he climbed in the water overflowed onto the floor.

ⓑ **What do you think caused the water to overflow?**

That moment, he had a flash of inspiration. He leapt out of the bath and ran naked through the marketplace shouting 'Eureka!', which means 'I have found it!'

Archimedes had realised that he could measure the **volume** of an object by measuring the amount of water it **displaces**. His body had displaced the water that had overflowed onto the floor.

Archimedes burst into the throne room. 'Your Majesty, have you got a lump of gold the same as the one you gave to the goldsmith? I have worked out a way of testing your crown.'

Archimedes placed the crown on the left-hand side of a balance and then put the lump of gold on the right-hand side. The scales balanced.

'So, my crown is all gold,' the King said, somewhat surprised.

ⓒ **Did the crown and the lump of gold have the same mass?**

Obtaining evidence

'Wait,' exclaimed Archimedes as he filled a large washbasin with water. The wise men of the King's court watched, thinking, 'What's he going to do now?'

Archimedes carefully lowered the crown into the basin and collected the water that overflowed in a larger basin. He then did the same thing with the lump of pure gold.

'It's pushed out less water than the crown!' cried the King.

d **Did the crown and the lump of gold have the same volume?**

Evaluating the evidence

'The crown is not made of pure gold,' went on Archimedes. 'If it was pure gold, it would have pushed out the same amount of water. The goldsmith is a thief. He has used some silver in the crown and kept some of the gold for himself.'

Archimedes knew that silver is lighter than gold. We say it is less **dense**. So, to make the crown the same mass, more silver was needed. This meant that it had a greater volume than the lump of gold. Archimedes knew this because it took up more space and displaced more water than the lump of gold.

'Send for the goldsmith! He is not going to get away with this one,' shouted King Hiero.

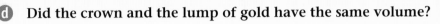

Mass, volume and density

Archimedes' test was based on **density**. The density of a lump of material depends on its mass and its volume.

$$density = \frac{mass}{volume}$$

The silver had the same mass as the gold but took up more space, because it has a lower density than gold. For example, 100 g of gold would take up 5.2 cm^3 but 100 g of silver would take up 9.5 cm^3 – almost twice as much volume!

Questions

1 Why do you think that the King was surprised when the crown had the same mass as the lump of gold?

2 Do you think that Archimedes would have come up with the answer if his bath had not been full to the brim? Give a reason for your answer.

3 Explain to your partner why the crown displaced more water than the block of gold.

4 What scientific knowledge did Archimedes use to explain his idea to the King?

5 Imagine Archimedes weighed the new crown and found that it weighed more than the original lump of gold. The goldsmith insisted that he had put extra gold in the crown to perfect the design.

   a What do you think Archimedes would have done next?

   b How do you think it would be possible for Archimedes to prove that the crown has some lighter metal in it?

6 How did he use his everyday experience to solve the problem of finding the volume of the crown?

L1 Shedding light

Seeing the light

Look at the photos below. All these objects produce light.

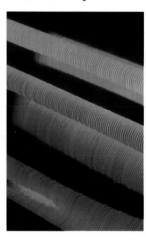  

a **Explain how we are able to see these things.**

Light travels in a straight line, at a speed of 300 000 000 m/s. It is sometimes difficult to understand just how fast light is. If you have ever been to America you will know that it takes a long time to fly there. Light is so fast that it gets there in a fraction of a second. In fact, it can travel from London to New York and back again more than 25 times in one second!

The Sun as a star

Our Sun is a source of light for everyone on Earth. It is a star, and all stars are sources of light. It takes just 8.3 minutes for the light from the Sun to reach us.

b **If it is 150 000 000 km from the Earth to the Sun and 5585 km from London to New York, how many times further is it from the Earth to the Sun than from London to New York?**

Like all stars, the Sun is a huge ball of very hot gas. The surface temperature of the Sun is 6000 °C.

This photo shows some stars in the night sky. Apart from our Sun, the nearest star is Alpha Centauri which is about 40 000 billion km (40 000 000 000 000 km) away.

Seeing stars

An orange fire is hotter than a red one. Some stars are red or orange, but the hotter ones are yellow or white. The Sun and stars are **luminous** which means they give out light. The **luminosity** of a star is a measure of the energy it gives out. The amount of energy depends on how big the star is and its temperature.

Colour	Temperature
red	3000°C
orange	4000°C
yellow	6000°C
white	10000°C
blue-white	20000°C

We can learn a lot about stars by looking at the light coming from them, as the table shows.

Light energy from the Sun is so strong that it masks the other stars, which are dimmer. So we can only see the other stars at night.

Most of the stars we can see belong to our galaxy, the Milky Way. A **galaxy** is a collection of millions of stars held together by gravitational pull. Our galaxy contains 100000 million stars. This photo of stars was taken by the Hubble telescope.

Seeing planets

Planets such as the Earth do not produce their own light. They are **non-luminous**. Look at the diagram. We see planets only because light from the Sun bounces off them (is **reflected**) and reaches us. We see the moons around the planets, including our Moon, because the Sun's light is reflected by them.

The further a planet is from the Sun, the less energy it receives from the Sun.

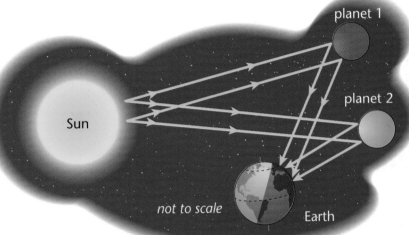

Sun

planet 1

planet 2

not to scale

Earth

Questions

1 Calculate how far light can travel in:

   a one minute   **b** one hour   **c** one day   **d** one year.

2 **a** The Sun is our nearest star, but it is so far away that it takes the light from the Sun about eight minutes to reach the Earth. Calculate how far away the Sun is.

   b The next nearest star is Alpha Centauri, 40000 billion km away. Explain why we only see the other stars at night, although we see our Sun during the day.

3 The Pole Star is one of the brightest stars that we see in the sky from the UK. Suggest why it is so bright.

4 Why can we see the Moon and other non-luminous objects in the Solar System, such as comets?

5 We can see some planets before it gets dark, such as Venus which is called the 'evening star'. We can see stars only at night. Why is this?

For your notes:

● The Sun and stars are **sources** of light.

● We see planets and moons because they **reflect** light from the Sun to us.

● Light travels in a straight line at 300000000 m/s.

● Most of the stars we can see belong to our **galaxy**, the Milky Way.

L2 All in a day

Learn about:
- Day and night
- Phases of the Moon
- Eclipses

Day and night

In the morning, the Sun appears to rise in the east. During the day, it appears to move across the sky. Then in the evening, it appears to set in the west. But the Sun does not move at all, it only seems to move because the Earth is rotating.

When the Sun shines on the rotating Earth, only the side of the Earth facing the Sun gets any light. The other half of the Earth is in shadow. As the Earth spins, the UK moves into the light. The rotating Earth gives us day and night.

Look at the diagram on the right. It is night in the UK.

a **Explain why the Sun 'appears' to move. What is actually happening?**

Through the night, the stars appear to move in a circle above our heads. To take this photo, the camera shutter was left open for four hours.

b **Why do the stars 'appear' to move across the sky?**

Do you remember?

The Earth spins on its **axis** once every 24 hours. The axis is an imaginary line that runs from the North Pole to the South Pole.

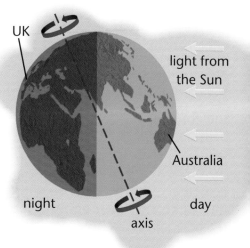

UK

light from the Sun

Australia

night   day

axis

Phases of the Moon

As the moon obits the Earth, its shape appears to change every day. These changes are called the **phases of the Moon**.

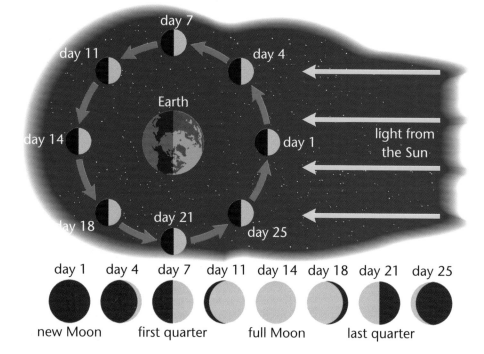

day 7

day 11   day 4

Earth

day 14   day 1

light from the Sun

ay 18   day 21   day 25

| day 1 | day 4 | day 7 | day 11 | day 14 | day 18 | day 21 | day 25 |

new Moon   first quarter   full Moon   last quarter

Do you remember?

The Moon also spins on its axis and it goes round the Earth once approximately every 28 days. The path it takes is called its **orbit**.

We see a full Moon when the Moon is on the opposite side of the Earth from the Sun. We see a new Moon when the Moon is between the Sun and the Earth, so there is no sunlight shining on the side we see from Earth.

The Moon's surface has craters. If you looked at its surface every night, you would always see the same craters. As well as orbiting the Earth in 28 days, the Moon also spins on its axis in 28 days. Try using an apple to model how this works.

Lunar eclipse

When the Earth is between the Sun and the Moon, we see a full Moon. Sometimes we also see a shadow of the Earth move across the Moon. This is called a **lunar eclipse**. Lunar eclipses can last for over an hour.

Lunar eclipses happen about twice a year. This is because the Moon's orbit is in a plane that is slightly tilted from the Earth's orbit. If both orbits were in the same plane, there would be a lunar eclipse every month.

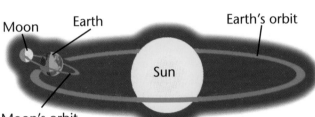

lunar eclipse – Moon in the shadow of Earth

Solar eclipse

Sometimes the Earth and Moon are in a position in their orbits where the Moon completely blocks the Sun's light from the Earth. This is a **solar eclipse**. The shadow of the Moon covers a small area of the Earth's surface in complete shadow. Here there is a **total eclipse**. Outside this area there is a **partial eclipse**. You can see a solar eclipse from only a limited part of the Earth, and it never lasts more than seven minutes.

Moon's orbit tilted at an angle from Earth's orbit.

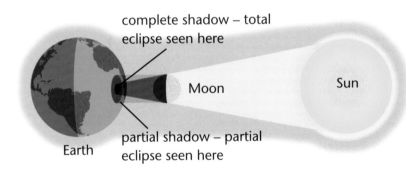

complete shadow – total eclipse seen here

partial shadow – partial eclipse seen here

Earth

Did you know?

The last total eclipse we saw in the UK was in 1999. The next one visible from the UK will be in 2090, but there will be others visible from other parts of the world before then.

For your notes:

- The Earth spins on its **axis** approximately once every 24 hours to give day and night.

- The Moon **orbits** the Earth once every 28 days to give the **phases of the Moon**.

- Sometimes when the Earth is between the Sun and the Moon, and there is a **lunar eclipse**.

- When the Moon blocks the Sun's light from reaching the Earth, there is a **solar eclipse**.

Questions

1 What would happen if the Earth were to turn more slowly on its axis?

2 Design some models using balls and lights that you can use with primary school pupils to teach them about:

   a night and day   b solar eclipses.

   Draw diagrams of how you would set up the equipment.

3 A lunar eclipse can last for over an hour, but a solar eclipse never lasts more than seven minutes. Explain why this is. Think about the relative sizes and distances of the Sun and Moon.

4 What would you have to do if you wanted to see a different part of the Moon's surface?

L3 All in a year

Learn about:
● Seasons

Earth years

The Earth actually orbits the Sun once every $365\frac{1}{4}$ days. Every four years we have a **leap year**. A normal year has 365 days, while a leap year has 366 days. The extra day comes from adding together the extra four quarters of a day every four years.

Do you remember?

As well as spinning on its axis, the Earth orbits the Sun once every 365 days. This is called a **year**.

Did you know?

The extra day is put in our calendar every four years on 29 February. If you were born on this day you would have a birthday only every four years!

The seasons

During the year, the climate in the UK changes. We have four **seasons**: spring, summer, autumn and winter. In the winter it is colder and in the summer it is warmer.

The axis that the Earth spins on is slightly tilted, at $23\frac{1}{2}°$ from the vertical. The UK is on the top half of the Earth, called the **northern hemisphere**.

When the northern hemisphere is tilted away from the Sun it is winter there, and when tilted towards the Sun it is summer.

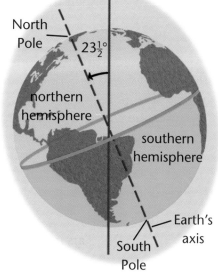

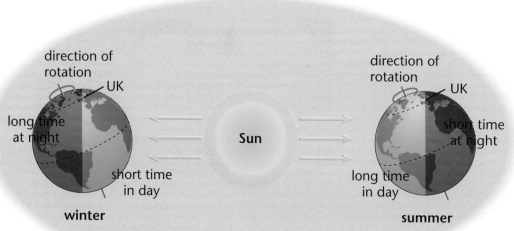

The tilt of the Earth means that in the summer the Sun seems high in the sky. In the winter the Sun seems lower in the sky. Because of the angle of the Earth to the Sun, different amounts of energy reach us in summer and in winter. In the summer, the high Sun shines directly on us, warming us up. In the winter, the low Sun's rays are spread over a large area, so it warms us up less.

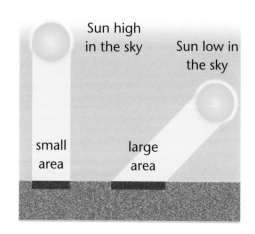

a It is hotter in summer and cooler in winter. What else changes about the days with the seasons?

In midsummer we have long days and short nights. The 21 June is called the **summer solstice**. This is the longest day, and the Sun is highest in the sky. In midwinter we have short days and long nights. The 21 December is the **winter solstice**. This is the shortest day, and the Sun is lowest in the sky. Spring and autumn have days and nights of equal length.

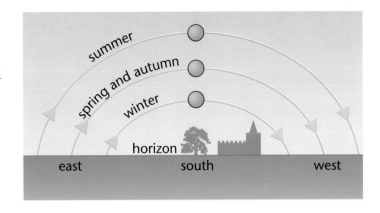

b Explain how the day length in the UK in summer and winter depends on the position of the Sun in the sky.

Life on the Equator

The **Equator** is an imaginary line that goes around the middle of the Earth, midway between the North and South Poles. The Sun is directly over the equator twice a year, on 21 March and 21 September. The climate at the equator does not change through the year very much, because the angle of the Sun's rays to this part of the Earth is less affected by the tilt of the Earth.

c The 21 March and 21 September are called the spring and autumn *equinoxes*. 'Equi' means equal and 'nox' means night. What do you think happens to the length of the day and night on these days?

Seasonal stars

Look at the diagram on the right. Imagine it is night as you look out at the stars in June and December. In June the UK is facing into space in one direction, and in December it is facing into space in the opposite direction. This means we see different stars. There are some stars and constellations we see all year round, but there are others we see only in the summer or winter.

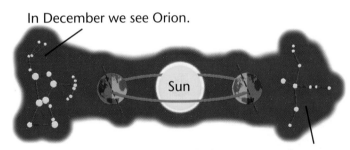

In December we see Orion.

In June we see Cygnus.

Questions

1 Make a list of all the changes between winter and summer that we notice in the UK.

2 Draw diagrams to show the shadow from a tree in the UK at midday on:

   a 21 June   **b** 21 December.

3 Australia is in the southern hemisphere. When the northern hemisphere is tilted towards the Sun, the southern hemisphere is tilted away from it. What would the climate and day length be like in Australia on:

   a 21 June?   **b** 21 December?

4 Imagine the Earth was not tilted on its axis. What would happen to:

   a the climate in the UK on 21 June?

   b the position of the Sun in the sky on 21 June?

For your notes:

- It takes the Earth $365\frac{1}{4}$ days or a year to **orbit** the Sun.

- The Earth's axis is slightly tilted at $23\frac{1}{2}°$ from the vertical. This causes the **seasons**.

- The stars we see in the night sky change with the seasons because we are facing a different way into space.

Orbiting the Sun

Our **Solar System** is made up of the Sun along with nine planets, including the Earth, orbiting the Sun. 'Sol' is Latin for Sun. The Solar System also contains other bodies measuring less than 1000 km across.

- **Asteroids** are mostly rocky objects found in the **asteroid belt** between Mars and Jupiter.

- **Comets** are clouds of gas with a bright centre of ice and dust particles and a tail.

- **Meteors** are solids objects that fall from outer space. A **meteorite** is a small meteor.

ⓐ What does the name 'Solar System' describe? Why is it a good name?

Planetary facts

The order of the planets
- Sun
- Mercury
- Venus
- Earth
- Mars
- Jupiter
- Saturn
- Uranus
- Neptune
- Pluto

Mercury is a small rocky planet that is closest to the Sun. It has no atmosphere and is very hot, up to 427 °C.

Venus is the hottest planet, about 480 °C. It is rocky, with an atmosphere of carbon dioxide.

Earth is a rocky planet with an average temperature of 22 °C. It has water and an atmosphere of nitrogen and oxygen. It is the only planet we know about with living organisms.

Jupiter is a cold (–150 °C), giant planet made mainly of liquids and gases. It is famous for its Great Red Spot, which is probably due to a giant storm. It has many moons.

Saturn is a cold (–180 °C), giant planet made mainly from gases. It has beautiful rings of ice particles around it and many moons.

Mars is a red, rocky, cold planet (–23 °C) with an atmosphere of carbon dioxide.

Uranus is a pale green, gas giant that has 15 moons and a number of rings like Saturn. The temperature is –210 °C.

Neptune is a bluish gas giant with a cold atmosphere (–220 °C). It has the strongest winds on any planet.

Pluto is rocky. It is the smallest and coldest planet (–230 °C) and is furthest from the Sun. Some people are not sure whether it is a planet at all because it is so small and has an irregular orbit.

ⓑ Suggest why it seems the Earth is the only planet suitable for life.

Planet years

It takes $365\frac{1}{4}$ Earth days, or a year, for the Earth to orbit the Sun. On other planets the length of a year depends on how far the planets are from the Sun and how long the orbits are. Mercury is closest to the Sun and takes only 88 Earth days to orbit the Sun. Pluto is the furthest from the Sun and takes 248 Earth years to orbit the Sun.

Collecting evidence

Astronomers have found out a lot about the Solar System using telescopes. In 1609 Galileo made the first telescope. This allowed him to see things in space magnified 30 times.

In 1770 Herschel, a church organist, was too poor to buy a telescope. He was so determined to look at the stars that he made a huge metal mirror in his kitchen and fitted it into a wooden tube eight feet long. He discovered Uranus with this telescope.

Later, when photography was invented, astronomers could take photographs through their telescopes. Pluto was discovered from a photograph in 1930. Modern telescopes like the Hubble telescope detect radio waves from distant stars and make pictures of them. These stars are too far away to send out visible light so we cannot 'see' them.

ⓒ Suggest reasons why Pluto was not discovered until 1930.

Did you know?

The Sun is much bigger than the Earth and the Moon is smaller than the Earth. It is only in the last 40 years that we have had photographs from space to show this.

Questions

1 Make up a sentence, using the first letter of each planet, to help you remember the order of planets in the Solar System.

2 We can see six planets with the naked eye. Which do you think they are?

3 Look at what the nine planets are made from. Can you divide them into groups? Explain your answer.

4 How many times longer does Pluto take to orbit the Sun than Mercury?

5 Discuss what would happen if the Earth were further away from the Sun.

6 a What is the relationship between a planet's distance from the Sun and its temperature?

   b Can you explain this relationship?

7 Describe two ways in which astronomers have used telescopes to find out more about the Solar System.

For your notes:

● The **Solar System** is made up of the Sun and the nine planets along with their moons and other objects such as **comets**, **meteors** and **asteroids**.

● The planets differ from each other in many ways, such as diameter, distance from the Sun, what they are made of and the conditions there.

131

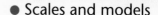
L5 Making models

The Solar System

Class 7J were studying the Solar System. They were drawing diagrams and making models that represented the Solar System. Some of the information they used is shown in the table. The Sun is 1 392 000 km in diameter. Look at the information carefully.

Planet	Diameter (to nearest 1000 km)	Average distance from the Sun (to nearest 1 000 000 km)	Furthest planet gets from Sun (to nearest 1 000 000 km)	Closest planet gets to Sun (to nearest 1 000 000 km)	Tilt of orbit (to nearest degree)
Mercury	5000	58 000 000	70 000 000	46 000 000	7
Venus	12 000	108 000 000	109 000 000	107 000 000	3
Earth	13 000	150 000 000	152 000 000	147 000 000	0
Mars	7000	228 000 000	249 000 000	207 000 000	2
Jupiter	143 000	780 000 000	816 000 000	714 000 000	1
Saturn	121 000	1 427 000 000	1 507 000 000	1 347 000 000	2
Uranus	51 000	2 871 000 000	3 004 000 000	2 735 000 000	1
Neptune	50 000	4 504 000 000	4 537 000 000	4 456 000 000	2
Pluto	2000	5 900 000 000	7 375 000 000	4 425 000 000	17

Relative sizes

Ian, Karl and Darren decided to make a model showing the sizes of the planets. Darren brought in a yellow beach ball to be the Sun. Karl and Ian brought in other balls to represent the planets.

Ball	Diameter in cm
beach	40
basketball	23.9
football	22.3
netball	21.3
volleyball	20.7
cricket	7.3
tennis	6.4
squash	4.4
golf	4.3
table tennis	3.8
marble	1.5

a Karl also brought in a rugby ball. Why did they decide not to use the rugby ball?

The table shows the balls in order of size.

b Decide which ball should stand for each planet.

Ian, Karl and Darren presented their model to the rest of the class. They said that their model shows the relative sizes of the Sun and the planets. The class **evaluated** the model. First, they said the good things about the model.

The balls are spheres, like the Sun and the planets.

Look back at both tables and compare the diameters of the balls with the diameters of the planets.

c Suggest ways in which this model represents the Sun and the planets.

Uranus and Neptune are almost the same size, and the netball and the volleyball are almost the same size.

The class then thought of ways to improve the model. Look back at the first table see what other information they had.

d Suggest how the model could be improved.

To improve on the model Ian, Karl and Darren decided to make their model to scale. They made a ball of clay to represent Pluto, the smallest planet. It was 2 mm in diameter.

e Using this scale, what will be the diameter of the Sun and the other planets?

Relative distances

Serena, Joyce and Parveen decided to make a model showing the distances of the planets from the Sun. They put labels on string to show the positions of the planets. They decided to use 1 mm of string for every 1 000 000 km, and to use the average distance of each planet from the Sun.

f Where should they put the labels for the Sun and each of the planets?

The class then evaluated the model.

g Use the comments on the right, and your own ideas, to write an evaluation of the string model. Check again what information they had. Include possible improvements.

Sizes and distances

Class 7J decided that they would build one model that shows both the sizes of the planets and their distances from the Sun. They decided to use the scale that Ian, Karl and Darren used: 1 mm = 1000 km.

h Using the average distance from the Sun, work out the distance to each planet in the model using this scale.

i Do you think it is possible to build a model that shows both the sizes of the planets and their distances from the Sun? Give reasons for your answer.

It shows how close together the inner planets are, and the big gaps between the outer planets.

The distances are to scale.

Sometimes the distances are shorter and sometimes they are longer.

Sometimes Pluto is inside Neptune's orbit.

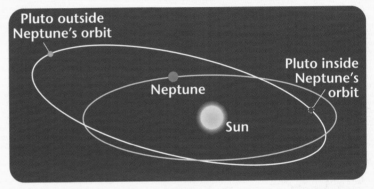

Pluto outside Neptune's orbit

Pluto inside Neptune's orbit

Neptune

Sun

It only shows the average distances.

It's 1D, and the Solar System is 3D.

It makes the planets look too close together. They are often on opposite sides of the Sun.

Questions

1 Look at the diagram of the Solar System on page 130.

2 Write an evaluation of this diagram as a model for the Solar System. Use the information in the table at the top of page 132 to make your decisions.

3 The diagram below shows a model of the inner planets of the Solar System.

    a Make an accurate copy of the diagram.

    b Use the information in the table opposite and label your diagram to show what each line represents.

    c Write an evaluation of this model.

Glossary

A The short way of writing amps.

acid A solution that has a pH lower than 7.

acid rain Rain polluted by acidic gases dissolved in it. Acid rain is more acidic than rainwater that is not polluted.

adaptations Having features that help a living thing to survive in a particular place.

adapted A well-adapted organism has features that help it to survive in a particular place.

adolescence The time in a young person's life when physical and emotional changes happen.

afterbirth The placenta comes out of the uterus after the baby is born. It is called the afterbirth.

air resistance The friction a moving object makes with air.

alkali A base that dissolves in water, forming a solution with a pH greater than 7.

alternative energy resources Energy resources that are not fossil fuels.

ammeter A device that measures the current in an electrical circuit.

amphibians One of the groups of vertebrate animals. Amphibians lay eggs in water but breathe air. They have a smooth, moist skin.

amps Current is measured in amps.

animal cells The building blocks that make up all animals. Animal cells have a cell membrane, cytoplasm and a nucleus.

animals Living things that feed on other living things and move around.

anther The part of the stamen in a flower that makes the pollen.

antibodies Substances in the blood and in breast milk, that protect the body from disease.

arthropods Group of invertebrate animals with segmented bodies and jointed legs.

asteroid belt A zone between Mars and Jupiter where most of the asteroids in our Solar System are.

asteroids Rocky objects in space. In our Solar System most of the asteroids are found in the asteroid belt.

atom The smallest part of an element.

axis An imaginary line through the Earth that runs from the North Pole to the South Pole.

balanced forces Two forces of the same size pulling in opposite directions.

base A substance that reacts with an acid and neutralises it.

battery Stores chemical energy. More than one cell connected together in an electrical circuit.

biomass The total mass of a living thing, not including the water. Biomass can be used as an energy resource.

birds One of the groups of vertebrate animals. Birds lay eggs with hard shells, look after their young and have feathers and wings.

braking distance The distance a car travels after the driver puts the brakes on but before it stops completely.

camouflage Features that help a living thing to blend in with its surroundings.

carbonates Substances that react with acid to produce carbon dioxide. Many rocks are made of carbonates.

carbon dioxide A gas that is produced when carbon burns and joins with oxygen, or when a carbonate reacts with an acid. Carbon dioxide turns limewater milky.

carnivore An animal that feeds on other animals.

carpel The female organ in a flower, that produces the egg cells.

cell (electrical) An object that changes chemical energy into electrical energy.

cells (in living things) Tiny building blocks that make up all living things.

cell division Cells split to make more cells.

cell membrane A thin layer that surrounds the cell and controls the movement of substances in and out of the cell.

cell wall A tough box-like structure around plant cells.

cellulose A tough stringy substance found in plant cell walls.

centipedes One of the groups of arthropods. Centipedes have lots of legs and a segmented body.

cervix A ring of muscle at the opening of the uterus.

charcoal A fuel made from wood, that is mainly carbon.

chemical change A change that makes a new substance. Many chemical changes are irreversible.

chemical energy Energy stored in a material, which will be given out in a chemical reaction.

chemical reaction A change that makes a new substance.

chlorides Substances that are formed when hydrochloric acid is neutralised.

chlorophyll A green substance that is needed for photosynthesis.

chloroplasts The parts of a plant cell that carry out photosynthesis.

chromatography A method used to separate mixtures of substances. The most soluble substances travel the furthest up the paper.

cilia Tiny hairs on the outside of some types of cell.

ciliated epithelial cell A specialised type of cell with cilia on its surface.

classification Putting things with similar features into the same group.

coal Material from plants that lived many millions of years ago, used as a fuel.

combustion The chemical reaction that happens when something burns.

comet A small object in the Solar System made of ice, dust and gas, that travels around the Sun.

compensation Making up for a change by balancing things out.

complete circuit Cells and lamps or other devices joined by wires to make a closed loop.

concentrated A solution that contains a lot of dissolved solute is concentrated.

condense Change from a gas to a liquid.

condition Something in a habitat that varies and can be measured, such as temperature or rainfall.

conserve/conserving Using fuels or other resources carefully, so that they will last longer.

conserved When the same amount is there at the end as there was at the beginning. For example, when a solid is dissolved in a liquid, the mass of solution equals the mass of solid plus the mass of liquid. The mass is conserved.

consumer An animal, that eats (consumes) plants or other animals.

contractions The muscles of the uterus wall squeeze when a baby is born.

cord This links the developing baby to the placenta in a pregnant female animal.

corrosion Eating away of the surface of a solid by a chemical reaction.

corrosive Substances that may destroy living tissues on contact are corrosive.

crustaceans One of the groups of arthropods. Crustaceans have lots of legs, a soft body and usually a hard shell.

current Electricity flowing around the circuit.

cytoplasm A jelly-like substance found inside cells.

deciduous Plants that are deciduous lose their leaves in the winter to become dormant and survive the winter.

dense A dense material has a lot of particles in a small volume.

density How heavy a material is for its size.

diffuses A gas or liquid spreads out and mixes with the gas or liquid around it.

diffusion Gas or liquid particles spreading out as their particles move and mix.

dilute A solution that does not contain much dissolved solute is dilute.

displaces Pushes out or replaces.

distillation A method used to separate the solvent from a solution, or to separate mixtures of liquids with different boiling points.

distilled water Water that has been made pure. It has been changed to a gas and condensed back to a liquid again.

diurnal An animal that is active during the day and rests at night is diurnal.

dormant An inactive state that allows an organism to survive harsh conditions, such as the winter.

dyes Coloured substances.

egg *or* **egg cell** The female sex cell in an animal or plant. The egg joins with the male sex cell in reproduction.

electrical energy Energy carried by electricity.

electron microscope A microscope that uses electrons instead of light. It makes things look very much larger.

element A substance that contains only one type of atom.

embryo A tiny ball of cells formed from the fertilised egg in animal reproduction.

embryo plant A new plant inside a seed ready to grow.

energy Energy makes things work. When anything happens, energy is transferred.

energy transfer The movement of energy from one place to another.

energy transfer diagram A diagram with arrows that shows how energy moves from place to place.

environment The surroundings.

environmental variation Differences in features that are affected by our surroundings are examples of environmental variation.

epidermis The outer tissue of human skin, or the upper and lower layer of cells in a leaf.

Equator An imaginary line that goes around the middle of the Earth, midway been the North Pole and the South Pole.

equinox A day of the year when day and night are the same length. There is one equinox in the spring and another in the autumn.

evaluate To judge how good a model or experiment is, finding its good points and bad points.

evaporation Changing from a liquid to a gas.

exoskeleton A hard outer coating that sometimes forms a shell. Arthropods have an exoskeleton.

expand To get bigger. A solid expands when you heat it because the particles move faster and take up more space.

extension The amount a spring stretches when you hang a weight on it.

fault Something that stops an electrical circuit working.

features Special parts of organisms, or particular things they do.

fertilisation In an animal, a sperm joining with an egg to make a baby. In a plant, a pollen grain joining with an egg cell to make an embryo plant.

fetus A developing baby inside the uterus of a female mammal.

filament A very thin wire inside a light bulb, that glows when the current passes through it.

fire triangle A way of showing the three things a fire needs to burn – fuel, oxygen and energy.

fish One of the groups of vertebrate animals. Fish live in water and lay eggs there. They breathe through gills and have scales and fins.

flatworms One of the groups of invertebrate animals. Flatworms have a flat leaf-shaped body.

float An object floats when the upthrust is equal to its weight. It stays on the top of the water.

food chain A diagram that shows how the organisms in a habitat feed on each other.

food web Two or more food chains link together to form a food web, that shows the feeding relationships between the organisms.

force arrows Arrows we draw that point in the direction of a force. The length shows the size of the force.

fossil The remains of an animal or plant that have been buried deep underground for millions of years and preserved.

fossil fuels Materials from animals and plants that lived many millions of years ago, used as a fuel.

friction The force that is made when things rub together.

fruit A structure made in a flower, that contains the seed. It is formed from the ovary.

fuel A material that has a lot of stored chemical energy. We burn a fuel to use the energy.

function The job that something does.

fungi Living things that feed on rotting material, for example, toadstools.

fuse A safety device for electrical circuits. The fuse has a very thin wire which melts if the current gets too high, or there is a fault in the circuit, and breaks the circuit.

galaxy A collection of millions of stars held together by gravitational pull.

gas A state of matter that is not very dense. A gas is easily squashed. Its shape and volume can change.

gas pressure A force caused by gas particles hitting the sides of their container.

generator A device that takes in kinetic (moving) energy and gives out electrical energy.

geothermal energy The electrical energy produced using the heat from the Earth's core.

gestation period The time a baby takes to develop inside its mother before it is born.

glands Parts that make hormones and other substances in animals. In male animals, the glands in the reproductive system make a liquid which mixes with sperm to make semen.

gravitational attraction/gravity The force that pulls everything towards the centre of the Earth. The other planets, the Moon and the Sun also pull things because of gravity.

gravitational energy Energy stored because something is lifted up.

greenhouse effect The carbon dioxide in the air stops some of the heat energy escaping from the Earth, making the Earth warmer. It behaves like the glass in a greenhouse.

growth Increase in size. Organisms grow by increasing the number of cells and increasing the size of the cells.

habitat The place where an animal or plant lives.

harmful Harmful substances may have a health risk similar to but less serious than toxic substances.

heat energy Energy transferred from a hot object to a cooler object.

HEP A short name for hydroelectric power.

herbivore An animal that feeds on plants.

hibernation An animal goes into a deep sleep to survive difficult conditions in the winter.

hormone A substance in the body that makes changes happen.

hydrocarbons Fuels that are made up of carbon and hydrogen.

hydroelectric power Using the kinetic (movement) energy of falling water to produce electrical energy.

hydrogen An element that is an explosive gas.

identical twins Two babies that came from the same sperm and egg. They are born at the same time and they look exactly the same.

implantation In animal reproduction, a fertilised egg settles into the soft lining of the uterus.

indicator A coloured substance that shows whether the solution being tested is acidic, alkaline or neutral.

infertility Not being able to reproduce naturally. If the man or woman is infertile, the couple cannot have babies without treatment.

inherited Passed on from the parents to their offspring.

inherited variation Differences in features that are passed on from the parents are examples of inherited variation.

input variable The thing you change in an investigation.

insects One of the groups of arthropods. Insects have six legs and a three-part body.

insoluble A substance that is insoluble will not dissolve.

interdependence Organisms in the same food chain all depend on each other.

intertidal area The area of beach that is under water at high tide but exposed at low tide.

invertebrates Animals without backbones.

irreversible change Something that cannot be changed back to how it was before.

irritant Substances that can cause redness or blistering in contact with the skin are irritant.

J The short way of writing joules.

jellyfish One of the groups of invertebrate animals. Jellyfish have a soft jelly-like body.

joules Energy is measured in joules.

kg The short way of writing kilograms.

kilograms Mass is measured in kilograms.

kilojoules There are 1000 joules in 1 kilojoule.

kilometres per hour Speed may be measured in kilometres per hour.

kinetic energy The scientific name for movement energy.

kJ The short way of writing kilojoules.

km/h The short way of writing kilometres per hour.

leap year A year that has 366 days, that occurs every four years.

lethal Something that is lethal can kill.

light energy Energy transferred by light.

light source Something that gives out light energy.

lime A basic substance containing calcium oxide, or other calcium compounds.

limewater A solution used to test for carbon dioxide. Limewater turns milky when carbon dioxide bubbles through it.

liquid A state of matter that flows. The shape of a liquid can change, but its volume is fixed.

litmus An indicator made from lichens. Acids turn blue litmus red. Alkalis turn red litmus blue.

lubricant A substance that reduces friction by making surfaces run smoothly against each other.

luminosity A measure of the light energy given out by bodies such as stars.

luminous Objects that give out light are luminous.

lunar eclipse An eclipse that occurs when the shadow of the Earth moves across the Moon.

m/s The short way of writing metres per second.

magnify To make something look bigger.

mammals One of the groups of vertebrate animals. Mammals have hairy skin. Their babies develop inside the mother and are fed on milk.

mammary glands Features that female mammals have, that make milk.

mass A measure of how much matter an object has.

material Anything that is made up of particles. A material may be a solid, a liquid or a gas.

matter Anything that has mass is made up of matter. Matter contains particles.

menopause Time in a woman's life when her periods stop.

menstrual cycle A monthly cycle in women. During the cycle an egg is released, and the woman has a period.

meteor A very small piece of debris from a comet.

meteorite A solid mass of rock or metal from space that lands on Earth. (A small meteor.)

methane A hydrocarbon fuel that is a gas. Natural gas is mainly methane.

metres per second Speed may be measured in metres per second.

microorganism A very small living thing that can only be seen with a microscope.

microscope A device that is used for looking at very small objects.

microscopic Something that can only be seen with a microscope is microscopic.

migration Moving to another habitat to avoid difficult conditions, for example, swallows fly south to avoid the cold winter in the UK.

millipedes One of the groups of arthropods. Millipedes have lots of legs and a segmented body.

model An idea or picture made up by a scientist to show a situation that cannot be seen. A model helps scientists think through explanations.

molecule A group of two or more atoms joined together.

molluscs One of the groups of invertebrate animals. Molluscs have a soft muscular body with a foot, and usually a hard shell.

movement energy When something moves, it has movement (kinetic) energy.

multicellular A living thing that is made up of more than one cell is multicellular.

N The short way of writing newtons.

natural gas A gas formed from animals and plants that lived many millions of years ago, used as a fuel. It is mostly methane.

neutral A substance that is neither acidic nor alkaline, with a pH of 7, is neutral.

neutralisation The chemical reaction that takes place when an acid reacts with a base.

newtons Force is measured in newtons.

nocturnal An animal that rests during the day and is active at night is nocturnal.

non-identical twins Two babies that came from different sperm and eggs. They are born at the same time, but look different.

non-luminous An object that does not give out light is non-luminous.

non-renewable energy resource An energy resource that cannot be replaced as we use it is non-renewable.

northern hemisphere The top half of the Earth, above the Equator.

nucleus The part of a cell that controls everything the cell does.

oestrogen A hormone in female animals that makes changes happen at puberty.

oil A liquid formed from animals and plants that lived many millions of years ago, used as a fuel.

omnivore An animal, that feeds on both plants and animals.

orbit The path a body takes around the object it is travelling round, such as the Moon's orbit around the Earth.

organ A group of different tissues that work together to carry out a function.

organism A living thing, that carries out the processes of life.

outcome variable The thing that changes during an investigation. The outcome variable is the thing you measure.

ovary In an animal, part of the female reproductive system that makes the eggs. In a plant, part of the carpel that makes the egg cells.

oviduct A tube in the reproductive system of a female animal. The eggs travel down the oviduct to the uterus.

ovulation An egg is released into the oviduct from the ovary.

oxide An oxide is made when a substance burns and joins with oxygen in the air.

oxygen An element that is a gas. Oxygen is used in burning and in respiration.

palisade cells The cells in a leaf where photosynthesis takes place.

palisade mesophyll The tissue in a leaf where the palisade cells are.

parallel circuit A circuit with more than one loop.

partial eclipse An eclipse viewed from a place where the shadow is not complete.

particle model The idea that everything is made up of particles.

particles Tiny parts that make up every type of matter.

penis Part of the reproductive system in a male animal. The penis allows the sperm to be placed inside the vagina.

period Part of a woman's menstrual cycle. The lining of the uterus breaks down and leaves the body through the vagina.

petal The part of a flower that is often colourful and attracts insects.

pH scale A number scale used to measure the strength of acidity and alkalinity.

phases of the Moon The different shapes of the Moon we see as the Moon orbits the Earth.

physical change A change in which no new substance is made. A change of state is a physical change. Physical changes are reversible.

placenta Structure formed in a pregnant female mammal. The developing baby gets its food and oxygen from the placenta.

plant cells The building blocks that make up all plants. Plant cells have a cell membrane, cytoplasm and a nucleus, and also a cell wall, chloroplasts and a vacuole.

plants Living things that are green and make their own food using sunlight.

pollen grains The male sex cells in a plant. A pollen grain joins with an egg cell to make the seed.

pollen tube A tube that grows from the pollen grain on the stigma, down the style to the ovule, so that the pollen grain nucleus can reach the egg cell.

pollination The transfer of pollen from an anther to a stigma in plant reproduction.

predator An animal that hunts other animals.

pregnancy The time when a female animal has a baby growing inside her uterus.

pregnant A female animal is pregnant when there is a baby growing inside her uterus.

prey Animals that are hunted and eaten by predators.

producer A plant, that produces its own food by photosynthesis.

products New substances that are formed in a chemical reactions.

puberty The first part of adolescence, when physical changes happen.

pure A pure material only contains one substance.

range The different values that are possible, such as all the different heights in a group of people.

reactants The substances that take part in a chemical reaction, that change into the products.

reaction force A force that stops things falling through solid objects. When you sit on a chair, your weight is balanced by the reaction force from the chair.

reflected When light bounces off a surface, it is reflected.

relationship A pattern that links variables together. A relationship describes how the outcome variable changes when the input variable is changed.

renewable energy resource An energy resource that can be replaced as we use it.

reproduction To make more organisms of the same species.

reptiles One of the groups of vertebrate animals. Reptiles breathe air and lay eggs on land. They have a scaly, dry skin.

resistance How much something slows down the electric current passing through it. A thin wire slows down the current more than a thick wire, so it has a higher resistance.

resultant force The size of an unbalanced force, which makes the object move or speed up or slow down.

reversible change Something that can be changed back to how it was before.

ring main The electrical circuit in a house. It is a parallel circuit.

roundworms One of the groups of invertebrate animals. Roundworms have a soft, thin round body.

sample A small part of something, used to represent the whole.

saturated A solution is saturated when no more of a solute can dissolve in it.

scale A scale drawing or model shows something bigger or smaller than it really is.

scale diagram A drawing that shows something bigger or smaller than it really is.

scale factor A number used in scale drawing. You multiply by the scale factor to scale something up. You divide by the scale factor to scale it down.

scaling down Making something smaller.

scaling up Making something bigger.

scrotum Part of the reproductive system in a male animal. The scrotum is a bag of skin that holds the testes.

seasons Times of different climate during the year. In the UK we have four seasons – spring, summer, autumn and winter.

seed A structure made in a flower, that contains the new embryo plant and a food store.

segmented worms One of the groups of invertebrate animals. Segmented worms have a soft ringed body.

segments Sections of the body in arthropods and segmented worms.

semen A mixture of sperm and a special liquid to help them swim.

series circuit A circuit in which everything is in one loop.

sexual intercourse The man's penis enters the woman's vagina, and sperm are released into the vagina.

sink An object sinks if its weight is bigger than the upthrust. It moves down in the water.

sodium chloride The scientific name for table salt.

solar eclipse An eclipse that occurs when the Moon blocks the Sun's light from reaching the Earth. A shadow passes across the Earth.

solar energy Energy given out by the Sun.

solar furnace A device that concentrates thermal energy from the Sun and uses the thermal energy to heat a material or to generate electricity.

Solar System The Sun and the objects orbiting it, including Earth and the other planets.

solid A state of matter that is dense and has a fixed shape and volume.

solubility A measure of how much of a solute will dissolve at a particular temperature.

soluble A substance that is soluble will dissolve.

solute The substance that dissolves to make a solution.

solution A mixture of a solute dissolved in a solvent.

solvent A liquid that substances can dissolve in.

sound energy Energy transferred by sound.

specialised A cell that is adapted to carry out a particular function is specialised.

species A particular type of animal or plant. Members of a species can reproduce to form more of their kind.

speed How fast something is moving.

sperm The male sex cells in an animal. The sperm joins with the egg in reproduction.

sperm tube A tube in the reproductive system of a male animal. Sperm swim from the testis to the penis through the sperm tube.

spiders One of the groups of arthropods. Spiders have eight legs and a two-part body.

stamens The male organs in a flower, that produce the pollen.

starfish One of the groups of invertebrate animals. Starfish have a hard star-shaped body.

stigma The part of the carpel where the pollen grain lands.

stopping distance The distance a car travels after the driver decides to stop but before it stops completely. Stopping distance = thinking distance + braking distance.

strain energy Energy stored in a material because the material is being pulled or pushed.

style The part of the carpel that holds up the stigma.

sublimation Changing straight from a solid to a gas without becoming a liquid.

sulphates Substances that are formed when sulphuric acid is neutralised.

summer solstice The longest day, on 21 June in the UK.

testis Part of the reproductive system in a male animal. The **testes** (plural) make the sperm.

testosterone A hormone in male animals that makes changes happen at puberty.

theory A set of ideas to explain something.

thermal energy The scientific name for heat energy.

thinking distance The distance a car travels after the driver decides to stop but before he or she puts the brakes on.

tissue A group of similar cells that carry out the same function.

total eclipse An eclipse viewed from a place where the shadow is complete.

toxic Toxic means poisonous. Substances that may cause serious health risks and even death if inhaled, taken internally or absorbed through the skin are toxic.

transferred Moved from one place to another.

twins Two babies that develop together inside the mother and are born at the same time.

unbalanced forces Forces pushing in different directions when one force is bigger than the other. An unbalanced force makes the object move or speed up or slow down.

unicellular A living thing that is made up of only one cell is unicellular.

universal indicator An indicator that has a range of colours showing the strength of acidity or alkalinity on the pH scale.

upthrust The force caused by water pushing up against an object.

uterus Part of the reproductive system in a female animal. The baby grows and develops in the uterus.

V The short way of writing volts.

vacuole A bag inside plant cells that contains a liquid which keeps the cell firm.

vagina Opening to the reproductive system in a female animal. Sperm enter the woman's body through the vagina, and the baby leaves through the vagina when it is born.

variable A thing that we change or that changes in an investigation.

variation The differences between living things, or between members of a species.

vertebrates Animals with backbones.

viruses A group of microorganisms that are not made of cells but reproduce inside other cells. Viruses cause disease.

voltage How much energy the electricity is carrying.

volts The energy a cell stores is measured in volts.

volume How much space something takes up.

water turbine A device that takes in the kinetic (movement) energy of falling water and gives out electrical energy.

water vapour Water that has turned into a gas.

wave energy The kinetic (movement) energy of waves.

wave turbine A device that takes in the kinetic (movement) energy of waves and gives out electrical energy.

weight The force of gravitational attraction on an object, that makes it feel heavy.

wind energy The kinetic (movement) energy of the wind.

wind turbine A device that takes in the kinetic (movement) energy of the wind and gives out electrical energy.

winter solstice The shortest day, on 21 December in the UK.

word equation An equation in words to show a chemical reaction.

year The time taken for the Earth to orbit the Sun.

Index

Note: page numbers in **bold** are for glossary definitions

Index